I0475661

INTRODUCTION

Milk has been rated as complete food for infants and a rich food supplement for the adults. It has been also recognized as the excellent food for the maintenance of the health and promotion of the growth of the human beings with the understanding of the nutritional aspects of various constituents of milk. The people are appreciating the value of the constituents of milk other than milk fat. Similarly, dairy technologists are engaged to see that each and every constituent of the milk is properly utilized.

India's milk production has reached to 84 million metric tones, (GOI, 2002), so is now looking for the new outlets for variety of dairy based convenience foods. The fluid milk consumption has become static. The product diversification is the need of the hour due to rapid changes in socio-economic status, living pattern and increase in urbanization. The dairy industries are also looking for the introduction of new products. (Singh *et al.*, 2002).

In India, the milk is consumed in the form of the fluid milk and various milk products i.e. butter, ghee, khoa, cheese, paneer, shrikhand, ice-cream etc. During manufacture of certain milk products, the by-products gets produced that too in very large amount than the products manufactured. The major by-products of the dairy industry are skim milk, butter milk and whey. Now a days more emphasis is given for economic utilization of these by-products as they contain valuable milk solids.

The utilization of these by-products has not only increased the availability of nutritional foods but has also indicated the profitable method of their utilization with minimizing the problem of pollution.

'Whey' can be defined as the opaque, greenish, yellow, watery fluid obtained as a by-product when milk is coagulated either by acid or rennett.

In India, whey is obtained as the by-product in the preparation of chhana, paneer, cheese, casein and shrikhand. Out of 85 million tones of global production 40 per cent is still disposed as raw whey in to sewage which leads to serious environmental pollution due to its high B.O.D. that is 3,00,000 – 5,00,000 ppm (Hofer, 1995). The whey which is used as the waste effluent could be used in the formulation of nutritious, palatable and therapeutic beverages.

On an estimate more than 3 million tones of whey is produced in country while more than 2 lakh tones of it containing valuable nutrients which are dumped in to gutter (Khamrui and Rajorhia, 1998). Ecologists are concerned about the heavy burden that whey makes on the sewage system and humanitarians feel that whey's valuable nutrients should be utilized in feeding the world's hunger.

A recent study indicates that treating 5 lakh litres of whey in sewage would cost $10,000 per day for primary treatment and $14,500 for tertiary treatment (Durham, *et al.* 1997).

There are different methods of processing of whey that is condensing, drying, fermenting but it has been found that these processes are not economical for utilization of whey, as these are of more energy consuming, costlier, high capital and labour demanding.

In India, cheese whey accounts for the major part of the total whey production i.e. nearly about 95 per cent. About 80 per cent of total whey produced is obtained from chhana, paneer and shrikhand production (Gupta and Mathur, 1989). In our country, 12 lakh tones of chhana is prepared per year, yielding about 8 million tones of whey (Aneja, 1997).

Cheese whey pollution is a major problem due to presence of high organic matter in the whey and its B.O.D. i.e. 36 mg/ml. (Shilpa and Gandhi, 2002).

Whey contains almost all the nutrients of milk except casein and fat. The whey has greater nutritional and therapeutic value because it contains whey proteins, lactose, thiamine, riboflavin, vitamin B6, Vitamin C,calcium and phosphorus (Belhe *et al.*, 1982).

Whey contains about half the total solids of milk (Gupta, 2000). Cheese whey contains about 20 per cent of the total milk proteins. (Khamrui and Rajorhia, 1998).

Whey also possesses preventive and curative elements and is especially used to treat a wide variety of aliments such as arthritis, anemia and liver complaints. These properties are additional virtues of the whey and should be exploited. (Jelen, 1992). Whey can also be used as an electrolyte solution (Gandhi, 1989).

The whey solids apart from being nutritious also carry excellent functional properties such as solubility, gel formation, emulsification, water binding, whipping etc. (Patel *et al.*, 1991). The whey solids are also rich in high quality proteins, minerals and easily digestible carbohydrates.

The whey solids can be potentially utilized for the production of lactose, alcohol, organic acid, protein recovery and as dried powder in various preparation (Gandhi, 1989). Whey can also be used in the preparation of bakery products, dried infant food, beverages, dried soup, frozen desserts, dry mixes etc. (Belhe *et al.*, 1982). Whey can be utilized in the preparation of ethanol, wine, acetic acid, dairy gels, lactose based sweeteners, whey syrups, crackers, vegetable soups etc. (Singh *et al.*, 1994).

There is an increased awareness all over the world on the potential utilization of whey primarily because of stringent pollution prevention regulations and secondly salvaging the unique components of the whey to ease the world food shortage.

Beverages are nourishing pleasant drinks that can be either alcoholic or non-alcoholic. Most dairy beverages are of non-alcoholic type although few forms of beverages may contain small quantities of alcohol (maximum 10 %) and carbon dioxide (Gupta and Mathur, 1989).

Beverages are consumed by the people of all age groups as they are nourishing, pleasant drinks that provides energy, water to digest the food, regulate body temperature, prevent dehydration, quenches thirst and removes physiological tension (Jandal, 1996).

The conversion of whey in to variety of beverages on a commercial scale has an economic advantage, as the whole quantity is being used and there are no problems of left over residues. It also improves the economic viability of dairy plant. (Shaikh *et al.*, 2001).

The pre-eminence of whey on an excellent beverage base has been recognized well due to its genuine thirst quenching, refreshing nature with some health benefits. These beverages can also offer good potential profit margins and are less acidic than fruit juices.

The variety of whey beverages consisting, plain, carbonated and alcoholic have been successfully developed and marketed all over the world, because it holds great potential for utilizing whey solids. In India, a

3

number of refreshing whey drinks including low cost 'Whevit' and 'Acidowhey' are in the market (Gupta, 2000).

Whey beverages already have achieved success in Europe as evidenced by the popularity of *'Rivellia'*, a deproteinized, fermented whey beverage from Switzerland; whey champagne and *'Kwas'* from Poland and *'Bodrost'* from U.S.S.R. Swiss scientists recently developed *'lacto fruit'* a non fermented, nutritious soft drink by using whey and partially hydrolyzed lactose syrup. (Fresnel and Moore, 1998).

Fruit whey drinks eg. *'Taksi'*, *'Yor'* are also enjoying commercial success in Netherland, chocolate drink based on acid whey *'Thumbs Up'* is widely marketed in U.S.A. Japanese product *'Milfull plain'* is also famous (Gupta and Mathur, 1989).

Currently, scientists developed the sport drinks vitamin fortified drinks, herbal extract beverages by utilizing the whey (Ghosh *et al.,* 1995).

A whey beverage can replace much of the lost organics and inorganics to extra cellular fluid. It is rapidly assimillable and forms an ideal metabolic substrate. From the aspects of necessity for health nutrition, whey possessed novel beverages acquired great importance.

Non alcoholic beverages such as fruit based drinks, synthetic drinks, carbonated drinks are in demand for great part of the year. The carbonated drinks i.e. cola type, flavoured with orange, lemon, mango have become popular.

Almost all the carbonated drinks contains synthetic colour and flavour and these are potentially allergenic. These soft drinks don't contain any fruit juice. Incorporation of the fruit juice in the soft drink not only improve its colour and flavour; but also provides nutrients and increases the nutritional value.

Beverages based on whey continue to receive a considerable amount of attention reflecting growing awareness of the potential of this product in the market place. These beverages have high nutritional quality and increased energy value. These could be particularly useful in places where there is a lack of food and improper nutrition leading to deficiency of certain nutrients.

With the growing awareness about the nutrition and health consciousness, the whey beverages also find its due place as refreshing and nourishing drink among the people. Thus, keeping nutritional and therapeutic status of whey in mind and growing global food shortages, the most ideal way is to use this by-product in palatable form.

India has not only made great progress in milk production but it has also emerged as top fruit producer in the world (FAO, 1995). As much as 25-30 per cent of total fruit produced in India gets spoiled due to lack of appropriate post harvest techniques. The formulation of a new product using suitable combination of whey and fruit juices in the beverage form would permit the economic utilization of whey and fruit juices.

Hence, by considering the market demands and consumer preference, conversion of whey in to beverages is one of the most important avenues for utilization of whey in human food chain.

Thus, keeping in view the tremendous demand of less costlier, nutritious and delicious diet beverages, it seemed worthwhile to undertake the present investigation to develop carbonated fruit flavoured beverage by utilizing acidic whey (shrikhand whey) and fruit juices with the following distinct and definite objectives.

Objectives:

1) To study Physico-chemical composition of acidic whey (shrikhand whey).
2) To standardize acidic whey (shrikhand whey) base for beverage.
3) To standardize the process for manufacture of carbonated fruit flavoured beverage.
4) To assess overall acceptability of the beverage.
5) To study physical properties, chemical quality and nutritional aspects of beverage.

REVIEW OF LITERATURE

The available literature on acidic whey (shrikhand whey) is scanty. Amongst published literature regarding the use of acidic whey (shrikhand whey) in preparation of beverages, very little information is available in this delicacy. Therefore, various aspects of beverages prepared from different types of whey are reviewed herewith.

2.1 Chemical composition of whey

Bambha *et al.* (1972) reported the average chemical composition of acid whey. According to them, an acid whey contained 93 per cent moisture, 0.1 per cent fat, 1 per cent protein, 5.1 per cent lactose, 0.7 per cent ash and 0.4 per cent lactic acid.

Singh and Mathur (1973) studied the chemical composition of the chhana whey. They reported that on an average, the chhana whey contained 93.09 per cent moisture, 5.08 per cent lactose, 0.37 proteins, 0.54 per cent fat, 0.53 per cent ash, 6.91 per cent total solids and 0.46 per cent acidity.

De (1974) has given the chemical composition of the acid casein whey. According to them, the acid – casein whey contained 93.1 per cent moisture, 0.1 per cent fat, 1 per cent protein, 5.1 per cent lactose and 0.7 per cent ash.

Kosikowski (1979) reported the composition of the fluid acid whey. The fluid acid whey was of 93.50 per cent moisture and of 0.40 per cent acidity.

Belhe *et al.* (1982) determined the chemical composition of cheese whey. They reported that the cheese whey contained water-soluble vitamins as thiamine, riboflavin, nicotinic acid, vitamin B6, pantothenic acid, vitamin C, calcium and phosphorus. They further stated that on an average cheese whey contained β-lactoglobuline, α-lactoalbumin, immunoglobulin and lactoferrin as 3.2, 1.2, 0.9 and 0.02 grams per litre of whey respectively.

Kulkarni *et al.* (1987) has given the chemical composition of chakka whey. According to them chakka whey contained 5.93 per cent total solids, 0.18 per cent fat, 0.65 per cent protein, 4.59 per cent lactose, 0.55 per cent ash and 0.9 per cent acidity.

Jayaprakasha *et al.* (1985) studied the variations in composition, pH and yield of the whey from cheese, chhana and acid casein. They reported the following mean composition of the whey from cheese, chhana and acid casein as 0.27, 0.51 and 0.14 per cent fat; 0.85, 0.39 and 0.97 per cent protein; 4.81, 5.05 and 5.08 per cent lactose; 93.0, 93.62 and 93.05 per cent moisture; 6.37, 6.37 and 6.95 per cent total solids; 6.69, 5.60, and 5.10 pH, 78.50, 84.75 and 82.68 per cent yield respectively.

Gagrani *et al.* (1987) determined the chemical composition of chhana whey. They reported that the chhana whey contained 93.05 per cent moisture, 4.98 per cent lactose, 0.58 per cent proteins, 0.6 per cent fat, 0.7 per cent ash, 6.951 total solids and 0.4 percent acidity.

Marwaha and Kennedy (1988) reported that cheese whey contained 3.3 to 6.0 per cent lactose, 0.3 to 0.7 per cent protein, 0.20 per cent fat, 5.6 pH and traces of salts.

Paul (1990) determined the following mean composition of whey from chhana, cheddar cheese, and acid casein as 93.0, 93.0 and 93.0 per cent moisture, 0.5, 0.3 and 0.10 per cent fat, 0.40, 0.90 and 1.0 per cent protein, 5.10, 4.90 and 5.10 per cent lactose, 0.40, 0.60 and 0.70 per cent ash, 0.20, 0.20 and 0.40 per cent acidity respectively.

Jadhav *et al.* (1991) studied the chemical composition of chakka whey. According to them, the chakka whey contained 4.66 pH, 0.78 per cent acidity, 0.35 per cent fat, 8.02 per cent total solids and 0.52 per cent total ash.

Singh *et al.* (1994) reported the chemical composition of the paneer whey as 6.15 per cent total solids, 0.1 per cent fat, 0.5 per cent proteins and 0.61 per cent ash.

Khamrui and Rajorhia (1998) has given the chemical composition of shrikhand whey. They reported that, shrikhand whey contained 6.96 per cent total solids, 0.38 per cent protein, 0.43 per cent fat and 0.94 per cent acidity.

Gupta (2000) described the chemical composition of acid whey. According to them the acid whey contained 7 per cent total solids, 0.1 per cent protein, 5.1 per cent lactose, 0.7 per cent ash and 0.4 per cent acidity.

Kaur *et al.* (2000) determined the Physico-chemical composition of the paneer whey. They reported that the paneer whey contained 0.20 per cent acidity, 5.06 pH, 1.0132 (g/c3) specific gravity and 26.2 (sec) viscosity.

Ganasekar and Balaraman (2001) studied the chemical composition of acid whey. They reported that acid whey contained 6.9 per cent total solids, 5.0 per cent lactose and 0.7 per cent ash.

Shaikh *et al.* (2001) determined the chemical composition of paneer whey. They reported that the paneer whey contained 93.60 per cent moisture, 0.52 per cent fat, 0.42 per cent proteins, 4.98 per cent lactose, 0.48 per cent ash and 0.44 per cent acidity.

Paul *et al.* (2002) described the chemical composition of chhana whey. According to them, chhana whey contained 6.68 per cent total solids, 0.25 per cent fat, 0.9 per cent protein, 0.54 per cent ash and 4.85 per cent lactose.

Sarvana kumar and Manimegalai (2002) reported that paneer whey contained 7 per cent total solids, 0.1 gram per cent acidity, 5.2 pH and 5.2 per cent total sugars.

Kumar *et al.* (2003) reported that acid whey contained 6.42 per cent total solids, 0.53 per cent protein, 4.4 per cent lactose, 0.6 per cent minerals and 4.7 pH.

Sangu (2004) studied the chemical composition of chhana whey. They reported that the chhana whey contained 0.5 per cent fat, 0.54 per cent protein, 4.02 per cent lactose and 0.6 per cent ash.

The foresaid literature indicates that whey is a good source of protein, carbohydrates and minerals. It has been effectively utilized for the preparation of beverages of high nutritional quality and of increased energy value.

2.2 Utilization of whey

The variety of approaches has been suggested for the proper utilization of whey in different forms and this is confirmed by a large number of patents and processes developed by the advanced countries.

The various options available for the potential utilization of the whey are depicted in Figure 1.

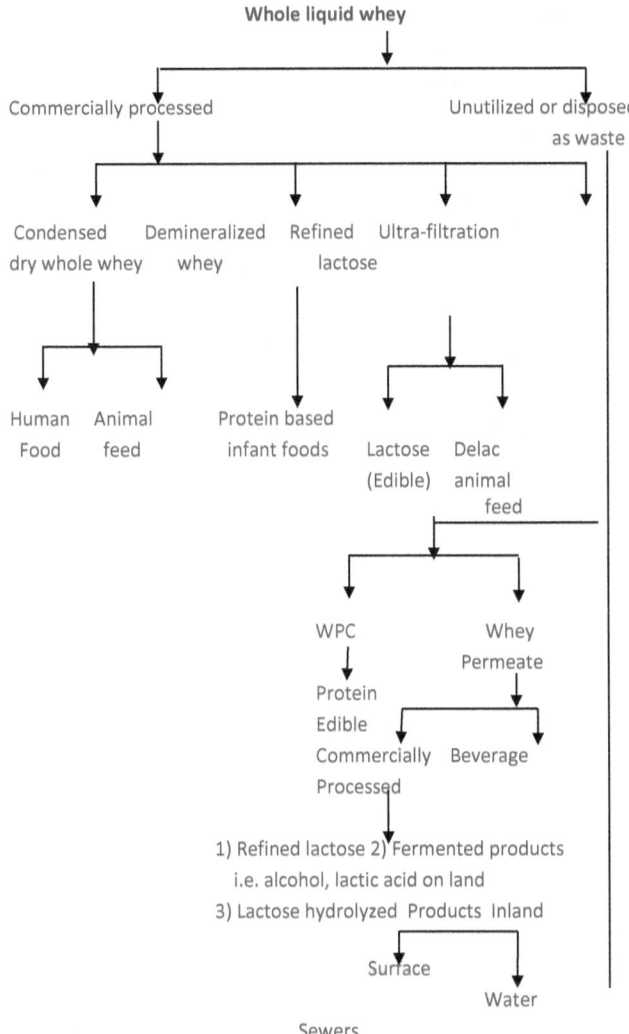

Fig. 1. Schematic presentation of whole whey processing
(Marhwaha and Kennedy, 1988)

2.2.1 Whey in human foods

Patel *et al.* (1991) discussed that whey can be converted in to a range of products i.e. whey powder, lactose, WPC, demineralized whey. These products can be used in infant foods, weaning foods, dairy products bakeries etc. They further reported the concentration of whey from 6 to 18 per cent total solids by reverse osmosis will save 45 per cent total energy and 25 per cent of over all cost.

Mohanty *et al.* (1998) discussed the various applications of whey for alcohol production acid beverages, sweet beverages, alcoholic and non-alcoholic beverages.

Gupta (2000) discussed the potential applications of the whey with respect to demineralization, ultra filtration, whey beverages and lactose production.

Ganasekar and Balaraman (2001) reported that whey and whey products can be used as milk replacer due to their high lactose content.

2.3 Beverages

2.3.1 Unclarified whey

Kosikowski (1968) produced an acceptable nutrient beverage by blending acid whey powder with frozen concentrated fruit juice, fresh fruit crystals and flavourings. Orange, grape fruit and pineapple juice nutrient beverages showed excellent flavour, texture and appearance qualities when 4 per cent acid whey powder was used compared to 6 per cent. Lemon juice, grape juice and flavoured powder nutrient beverages scored lower due to sedimentation.

Nelson and Brown (1971) prepared an orange flavoured drink containing 33 per cent unmodified cottage cheese whey, which was rated as 6.3 on 7 point hedonic scale compared to 4.7 of non whey drink. The whey contributed significantly to the colour and flavour stability of the drinks. They further prepared acceptable drinks of 80 – 90 per cent whey flavoured with 10 per cent strawberry puree or 20 per cent peach puree and 7 – 20 per cent grape fruit juice.

Bambha *et al.* (1972) manufactured 'Whevit' a nourishing soft drink, the orange, pineapple, lime juice and mango flavours were added to chhana whey. The manufacturing process comprised of steaming of whey, cooling, filtering, addition of sugar syrup and yeast culture. Then incubated and added with flavour, citric acid, colour, followed by bottling and pasteurization and stored at low temperature.

Nelson et al. (1972) suggested that orange flavour can be successfully used in the drink containing 75 – 90 per cent whey. They reported that synthetic red rasberry gave better results than some other colours. In the final product the brix was adjusted to 18 – 20 per cent and pH to 3.6 by adding sugar and citric acid respectively.

Holsinger et al. (1974) reported that the acid whey flavour was most compatible with citrus flavours particularly orange. Good consumers acceptance was claimed for orange flavoured acid whey beverage. The final product contained 0.7 to 1 per cent protein and could be carbonated.

Schuster (1977) patented the process for producing the beverage from acid whey. It was free from alcohol, CO_2 and preservative. The beverage comprised of 70 – 80 per cent whey, less than 5 per cent lactic acid, less than 5 per cent yoghurt, less than 5 per cent lactose and milk proteins. To this fruit juice and pulp were also added. The pH of final beverage was 4.2 and was stared for 6 – 7 weeks at 20°C or more than 2 months at 8°C without any preservative.

A variety of fruit whey beverages was prepared under the brand name 'trimo' in small packs. The flavours used were of black current, orange, passion fruit and mixed fruit. These were prepared by mixing drinking whey and fruit juices. (Anon, 1982).

Niketic and Marinkovik (1984) used acid whey, different stabilizers and fruit concentrates of orange mandarin, apple, lemon (2 – 4 %) to prepare soft drink. It was then sterilized, homogenized and packed in tetra packs. The shelf life of drink was 6 weeks at room temperature.

Barbier and Rialand (1987) described the process for preparing flavoured beverage using acid whey. The method comprised of acidification, mineralization of the whey by cation exchange, sterilization and sweetening and flavouring.

Rajeshkumar et al. (1987) developed lassi type cultured beverage from cheese whey by adding sugar, pineapple flavour and skim milk (95.5) with 12.5 per cent sugar. The synthetic pineapple flavour was found best.

Khalikar (1990) studied the utilization of acid whey for preparation of soft beverage. The final product contained 24 per cent sugar and 10 per

11

cent orange juice. It was then concentrated by thermal heating i.e. 70°C for 20 minutes resulting in to formation of quality beverage.

Singh *et al.* (1994) developed acceptable whey based mango, pineapple, lemon and banana beverages. Among these beverages, mango beverage was found to be preferred most with 15 per cent pulp, 78 per cent paneer whey, 7 per cent cane sugar and 4.5 pH.

Mandal *et al.* (1997) reported the preparation of beverage from chhana whey with the use of citric acid, sugar and lemon juice. The beverage with 10 per cent lemon juice, 75 per cent chhana whey and 8 per cent sugar was found to be preferred most. The Sorbic acid used was an effective preservative to extend shelf life.

Khamrui and Rajorhia (1998) manufactured ready to serve whey based kinnow juice beverage. The formulation containing 40 per cent juice, 53 % whey, 7 per cent sugar, 0.05 per cent pectin, 0.15 per cent CMC and 4.25 pH was found to be preferred mostly.

Krishnaiah *et al.* (1998) developed an acceptable quality beverage with good organoleptic qualities from acid whey; by mixing 3 parts of whey with 1 part of toned milk, to which 10 per cent sugar and pineapple essence was added. The pH of the final beverage was 6.0.

Kaur *et al.* (2000) developed whey based carrot juice beverage by incorporating four levels of carrot juice i.e. 5, 15, 25, 35 per cent in the whey to which 6.5 per cent sugar and 0.5 per cent salt was added. The sodium benzoate was added at the rate of 120 ppm. The whey beverage with 25 per cent carrot juice gave the best results.

Prasad *et al.* (2001) developed the whey based mango beverage. The specific proportion of sugar and acid were designed using "Central Composite Rotatable Design". The beverage containing 12 per cent sugar and 20 per cent whey resulted to 17°Bx and 48.57 to 51.52 total soluble solids acid ratio were found to be best.

Dhaka *et al.* (2002) prepared a whey based coffee drink. In 200 ml of drink, 120 ml of whey, 0.5 g of coffee, and 15.3 g of sugar was found most appropriate.

Sarvanakumar and Manimegalai (2002) developed the whey based jack fruit RTS beverage. The final product with 10 per cent pulp, 50°Bx and 0.3 per cent acidity was found to be most acceptable.

12

Sarvanakumar and Manimegalai (2003) prepared the whey based pineapple juice RTS beverage. The end product with 10 per cent juice, 15°Bx and 0.3 per cent acidity had highly acceptable taste and overall acceptability.

Shukla *et al.* (2004) conducted a study on the paneer whey beverages prepared by blending juice of apple, banana, guava, litchi and mango at four different levels i.e. 10, 20, 30 and 40 per cent. The organoleptic evaluation showed that acceptable quality beverage can be prepared with 20 per cent apple juice, 30 per cent litchi juice, 20 per cent banana pulp, 10 per cent guava juice and 30 per cent mango juice with 10 per cent sugar.

2.3.2 Clarified whey

Fresnel and Moore (1978) developed a non-fermented nutritious soft drink i.e. 'Lactofruit'. It was prepared from partially hydrolyzed lactose syrup. The hydrolysis was carried out by an immobilized lactate enzyme and ultrafiltered whey. The product obtained was clear, yellow coloured to which flavourings and sucrose were added.

Fauquant *et al.* (1985) reported a pilot plant process for clarification of acid whey by adjusting the pH to 7.2 and heating at 45°C for 20 min and centrifuging to remove precipitated proteins.

Jayaprakasha *et al.* (1985) reported the effect of chemical agents on clarification and deproteinization of different types of whey. The cheese whey, chhana whey and acid whey were treated with various concentration of bentonite for 3 hrs, keiselguhr for 2.5 hrs and fuller's earth for 0.5 hr prior to filtration and used in the preparation of soft drinks. The concentrations of chemicals required to eliminate turbidity from all types of whey was 6, 3 and more than 6 % bentonite, 4,3,6 % keiselguhr, 3, 2, 4 per cent fuller's earth.

Prendergast (1985) prepared a fermented fruit drink by using coarse filtered sour cheese whey. The beverage was inoculated with culture to get the optimum pH in the range of 3.8 to 4.2 followed by pasteurization then cooled to 42°C and the final beverage contained 0.30 percent stabilizers, 2 to 7 percent fruit concentration, 85.70 percent sour whey,6 to10 percent sugar with colour and flavour.

Albrecht (1986) reported two types of whey drinks i.e. apple and lemon by using filtered whey. The final beverage was pasteurized at 75°C then pH was adjusted to 3.5. Both drinks had good organoleptic properties and homogeneous consistency.

Mathur *et al.* (1986) described the various methods for clarification of the whey. The cheese, casein, paneer whey clarification was done by heating at 98°C for 10, 20 and 30 minutes respectively then adjusted pH value 3.5, 4, 4.5, 5 and 5.5 with or without addition of 0.2 to 0.5 per cent $CaCl_2$. The clarity of whey was much superior when $CaCl_2$ in the range of 0.2 – 0.5 % at pH 5.5 was added. They further stated that sodium hexametaphosphate helps in clarification of whey without application of heat and also preserve the nutrients.

Gagrani *et al.* (1987) prepared fruit flavoured beverage by using deproteinized and partially clarified chhana whey with fruit juices i.e. orange, pineapple, guava and mango. The final product was adjusted to 0.5 per cent acidity and 20°Bx. The beverage prepared from 15 per cent mango juice was superior in quality.

Reddy *et al.* (1987) prepared a sterilized whey beverage containing deproteinized cheddar cheese whey, lemon juice and sugar. Three different types of beverages were formulated using 6 per cent, lemon juice and 11 per cent sugar, 8 per cent lemon juice with 14 per cent sugar and 0.3 per cent lemon essence with 9 per cent sugar. The study indicated that acceptable quality beverage could be manufactured by using the combination of 8 per cent lemon juice and 14 per cent sugar.

Mathur *et al.* (1988) produced fruit whey drink using filtered acid whey with stabilizers. The beverage was fermented type and fruit concentrates was also added. It was also subjected to homogenization and UHT sterilization. The beverage was packed aseptically and found that it could be stored up to 4 months at 8 + 1°C and up to 1 month at 37°C.

Krishnaiah *et al.* (1989) developed the beverage from deproteinized acid whey. Three different types of beverages were formulated by using 10 per cent citric acid, 0.2 per cent acid whey and orange essence. Three parts of acid whey with one part of toned milk, 10 per cent sugar and pineapple essence. Three parts of acid whey and one part of toned milk with 10 per cent sugar and banana essence with lemon yellow colour. The final product was added with essence for pleasing purpose.

Krishnaiah *et al.* (1991) developed three types of whey beverages using acid whey; of these one was with addition of 10 per cent sugar, 0.21 acidity in deproteinized acid whey. The remaining two were prepared by mixing three parts of acid whey with one part of toned milk using different colour and flavour. All these three beverages were acceptable and had almost identical organoleptic qualities.

Shitova *et al.* (1991) patented a process for preparation of whey beverage that involves thermo chemical precipitation of whey proteins at $93 - 97^{\circ}C$, cooling, acidification to pH $3 - 4.5$ and separation of resultant mixture in to a protein component and clarified whey.

Jayaprakasha *et al.* (1992) prepared a whey drink by using deproteinized and clarified cheddar cheese whey with 10 per cent sugar, 0.4 per cent citric acid with mango, orange and pineapple flavourings.

Sikder (2001) clarified the chhana whey and prepared a whey based mango beverage. The chhana whey was clarified by passing through sterilized cotton pad and then deproteinized. The carbonated beverage with 12 per cent mango pulp and 8 per cent sugar was selected for product formulation.

Shaikh *et al.* (2001) clarified the whey by heating at $98^{\circ}C$ for 10 minutes at pre-adjusted pH of 4.5. It was then cooled to room temperature and kept for $5 - 6$ hrs to allow the precipitated proteins to settle down. The clarified whey was then passed through muslin cloth and cotton pad. Sugar level of 12 percent, 0.1 per cent pineapple flavour, carbonation at 72.48 kg cm^{-1} were found to be most acceptable.

Suresha and Jayaprakasha (2003) suggested ultra filtration for clarification of casein whey and permeate so obtained from ultra filtration was used for preparation of fruit flavoured beverage with pineapple mango and orange flavours by adding sugar, citric acid, colour and flavour. The beverage containing 10 per cent sugar, 0.1 per cent acidity was adjudged best compared to other combinations.

Suresha and Jayaprakash (2004) suggested the ultra filtration for the clarification of cheddar cheese whey. They prepared a fruit flavoured beverage from lactose hydrolyzed whey permeate with the addition of pineapple, mango and orange juice. The beverage containing 8 per cent sugar, 0.1 per cent citric acid and 20 per cent pineapple juice was adjudged as best.

2.4 Carbonation of whey beverages

Skudra and Reinikova (1976) reported a carbonated beverage with CO_2 concentration of 1 g per 500 ml, which completely inhibited the growth of moulds and yeast. A bacteriostatic effect was being observed when the CO_2 concentration was raised to 3-4 g per 500 ml.

Karunakar *et al.* (1984) demonstrated the preservative effect of solid carbon dioxide gas on the quality of the beverage. They found that solid CO2 was suitable for preserving the beverage.

Choi and Kosikowski (1985) made sweetened plain and strawberry flavoured carbonated beverage. The process comprised of blending 12 per cent sucrose and strawberry flavoured extract. It was carbonated under 0.5 kg /cm2 at 4°C.

Jayaprakasha *et al.* (1986) developed a soft drink by deproteinized and clarified cheese whey and chhana whey with the addition of 6-12 per cent sugar, 0.2-0.4 per cent citric acid, 0.15-0.45 ml/lit flavourings. The drink was filled in bottles, pasteurized, cooled and saturated with CO_2. The pasteurized, carbonated drinks were found to be most acceptable and had the highest mean sensory score of 90 (Max. 100).

Khurdia (1989) prepared carbonated fruit flavoured beverages. He stated composition of fruit based carbonated beverage. According to him, lime beverage of 10°Bx was carbonated at 100 psi, phalsa beverage of 15°Bx was carbonated at 80 psi, jamun beverage of 10°Bx was carbonated at 80 psi, dried ber juice of 9°Bx was carbonated at 120 psi, apple beverage of 14.5°Bx was carbonated at 100 psi and orange beverage of 10.4°Bx was carbonated at 100 psi.

Khurdia (1990) prepared fruit juice based carbonated drinks from lime, phalsa, jamun, ber and apple. The carbonated drink made from lime having 50 BAR (Brix – Acid Ratio) at 100 psi CO_2 from jamun, 28.57 BAR at 80 psi CO_2 from phalsa, 25 BAR at 80 psi CO_2 from ber, 45 BAR with 33 per cent dried ber juice at 120 psi CO_2 and from apple 41.4 BAR with 36 ml concentrated at 100 psi CO_2 were found to be the best.

Wazir Singh *et al.* (1999) standardized the technology for the manufacture of guava whey beverage using 1 : 3 guava extract and paneer whey with 8 per cent sugar with carbonation of final product.

Shaikh *et al.* (2001) reported the development process for preparation of fermented carbonated whey beverage. The beverage was

16

carbonated to 63.42 kg/cm^2, 72.48 kg/cm^2 and 81.54 kg/cm^2. Finally the carbonation of 72.48 kg/cm^2 was found to be most acceptable.

Suresha and Jayaprakasha (2003) studied the utilization of ultrafiltered whey permeate for the preparation of beverage. The permeate beverage after chilling at 4 $\pm$ 1°C was carbonated at 140 psi.

Suresha and Jayaprakasha (2004) studied the utilization of ultrafiltered lactose hydrolyzed cheddar cheese whey permeate for the preparation of fruit flavoured beverage. The permeate beverage after chilling at 4 $\pm$ 1°C was carbonated at 140 psi pressure.

2.5 Overall acceptability of beverage

Gagrani *et al.* (1987) added different fruit juices like orange, pineapple and mango at the rate of 10 per cent, 15 per cent and 15 per cent respectively to chhana whey for the manufacture of fruit based beverage. He reported that 15 per cent mango based beverage was superior than others with respect to colour, flavour, mouthfeel, and physical characteristics like sedimentation, turbidity and viscosity etc. The highest score was recorded for mango flavoured whey beverage (7.42) followed by pineapple flavoured whey beverage (6.97) and orange flavoured whey beverage (6.92).

Rajesh Kumar *et al.* (1987) developed a lassi type cultured beverage from cheese whey.They used different sugar concentration such as 10, 12.5 and15 percent and different synthetic flavours such as pineapple, rose and banana. The final product with 12.5 per cent sugar and synthetic pineapple flavour was "liked extremely" with highest score of 7.71 for pineapple flavour.

Reddy *et al.* (1987) studied the consumer's acceptance of lemon flavoured cheese whey beverage. The beverage with 8 per cent lemon juice and 14 % sugar stored at refrigeration temperature was preferred over the beverage with 6 per cent lemon juice and 11 per cent sugar stored at room temperature on the basis of sensory score.

Krishnaiah *et al.* (1989) developed three categories of acid whey beverages. The first category was made by the addition of 10 per cent sugar and 0.2 per cent citric acid to deproteinated acid whey with orange essence. The second category was prepared by addition of three parts of acid whey, one part of toned milk 10 per cent sugar with pineapple essence and yellow colour. The third category was prepared by mixing three parts of acid whey, one part of toned milk and 10 per cent sugar with banana essence and lemon yellow colour. The three whey beverages were acceptable with almost identical organoleptic qualities. The second and third categories of the beverages were most acceptable than first category due to added toned milk.

Singh *et al.* (1994) developed the panner whey based beverages by using mango, pineapple, lemon and banana juice. The beverage prepared from mango contained 15 per cent pulp, 78 per cent whey, 7 per cent sugar, the pineapple beverage was prepared by adding 20 per cent juice, 73 per cent whey and 7 per cent sugar, lemon beverage contained 5 per cent juice, 87 per cent whey and 8 per cent sugar, banana beverage contained 20 per cent juice, 73 per cent whey and 7 per cent sugar. The mango beverage showed highest (7.8) flavour score where as pineapple beverage possessed least score (7.2). The body score was highest (7.7) for mango beverage and was lowest for pineapple beverage (6.6). The mango beverage was also judged as highest (7.7) for colour. The overall acceptability of mango beverage was 7.7 followed by lemon beverage that is 7.5, banana beverage 7.3. The pineapple beverage was least (7.0) preferred beverage.

Khamrui and Rajorhia (1998) developed ready to serve whey based kinnow juice beverage. The various proportions tried included juice, whey and sugar content of the formulated beverage ranging from 15 – 50 per cent, 42 – 79 per cent and 6 – 8 per cent respectively. The formulation containing 40 per cent kinnow juice, 53 per cent whey, 7.1 sugar, 0.05 per cent pectin, 0.15 per cent CMC and pH 4.25 was found to be liked extremely by the sensory panel.

Wazir Singh *et al.* (1999) standardized the technology for the manufacture of guava whey beverage using three different ratios of guava extract (1:5, 1:4 and 1:3) and three edible colours (rose, orange, lemon) and four levels of sugar (5, 7, 8 and 10 per cent), The guava whey beverage with extract at 1 : 3 ratio with 8 per cent sugar and lemon colour was scored highest (7.58) compared to other formulations.

Kaur *et al.* (2000) developed whey based carrot juice beverage by using carrot juice at four different levels i.e. 5, 15, 25, 35 per cent in to paneer whey to which 6.5 per cent sugar and 0.5 per cent salts has been added at rate of 120 rpm. The incorporation of 25 per cent juice gave the highest sensory score (7.74) followed by 35 per cent juice (7.37), 15 per cent juice (7.18) and 5 per cent juice (6.04). The incorporation of 25 per cent juice gave the best results.

Prasad *et al.* (2001) recorded the overall acceptability of whey based mango beverage on a 9 point hedonic scale. According to them flavour and overall acceptability score for mango beverage varied from 6.23 to 8.76 and 6.96 to 7.95 respectively. The beverage containing 12 per cent sugar, 20 per cent whey were the best in the terms to their flavour and overall acceptability.

Shaikh *et al.* (2001) compared the market sample with the sample prepared in the laboratory of fermented carbonated whey beverage. The sensory evaluation was carried out on 9 point hedonic scale. According to them the highest score of 9 each was recorded with respect to taste, aroma, mouth feel and the products were 'liked extremely' in comparison with market sample. The score for consistency of market sample and developed beverage were observed to be 7.5 and 8.0 respectively and the slightly highest score to beverage may be due to its turbid nature which was 'liked very much' by the judges. The overall acceptability of beverage under study was also better than market sample.

Sikder *et al.* (2001) studied the overall acceptability of whey based mango beverage. The whey beverage formulated with 12 per cent juice and 8 per cent sugar scored highest. The carbonated sterilized whey beverage showed better acceptability than whey beverage prepared with preservative. The carbonated sterilized beverage stored at 30ºC $\pm$ 1ºC counts organoleptically in good condition up to 70 days where as sterilized beverage at 7 $\pm$ 1ºC was found to be organoleptically fit up to 80 days storage.

Sarvana Kumar and Manimegalai (2003) studied the overall acceptability of whey based pineapple juice RTS beverage on 4 point hedonic scale for a period of 3 months. The beverage had highly acceptable taste, colour and flavour up to 15 days and scored 4.0. The score was found to be decreased to 3.4 due to change in taste between 46 – 90 days. The highly acceptable status was noticed upto 45 days with highest score values (4.0) and then changed to 3.5 between 46-90 days.

Suresha and Jayaprakasha (2003) studied the overall acceptability of whey permeate carbonated beverage on 9 point hedonic scale. The carbonated beverage was served along with non carbonated beverage. According to them the average scores for flavours attribute was found to be 8.2, 8.15 and 8.10 respectively for pineapple, mango and orange flavoured beverage. The simillar trend for colour and appearance attribute was observed. The score awarded for colour and appearance attributes were only 8.23, 8.10 and 8.0 respectively. The consistency score before carbonation was found to be 8.12, 8.12 and 8.13 which upon carbonation improved to 8.40, 8.43 and 8.42 respectively. The overall acceptability score improved from 8.30 to 8.65, 8.10 to 8.40, 8.45 to 8.65 respectively for pineapple, mango, orange beverages. It indicates that carbonation improves colour, appearance, taste, consistency and overall acceptability.

Suresha and Jayaprakasha (2004) discussed the overall acceptability of beverage prepared from lactose hydrolyzed whey permeate on 9 – point hedonic scale. The scores with respect to hydrolyzed and non hydrolyzed beverage were found to be 8.55, 8.20, 8.50 and 8.15, 8.50, 8.10 for pineapple, mango, orange flavoured beverage respectively. The average score for overall acceptability of control and hydrolyzed beverages were 8.20 and 8.55, 8.10 and 8.45, 8.05 and 8.40 respectively for pineapple, mango, orange flavoured beverage. The significant improvement in overall acceptability scores of hydrolyzed beverages as compared to unhydrolyzed beverage could be attributed to the improvement in the scores for flavour, resulted in improving overall acceptability. The respective scores of carbonated and non carbonated hydrolyzed permeate beverage for colour and appearance were 8.65 and 8.25, 8.50 and 8.11, 8.55 and 8.15 respectively for pineapple, mango, orange flavours. The average scores awarded for flavour were 8.5 and 8.85, 8.50 and 8.80, 8.50 and 8.80 respectively. The scores for the overall acceptability of hydrolyzed non carbonated permeate beverage were observed to be 8.55, 8.45 and 8.45 respectively. Upon carbonation, the scores increased to 8.85, 8.70 and 8.70 respectively.

2.6 Physico-chemical and nutritional quality of whey beverage.

Bamha *et al.* (1972) discussed the Physico-chemical and nutritional qualities of 'Whevit' a nourishing soft drink. According to them, 'Whevit' contains 0.4 – 0.6 per cent protein, 10 – 11 per cent total sugars, 0.05 per cent fat, 0.5 – 0.7 per cent alcohol, 7 – 7.5 per cent acidity and 4.5 – 6 per cent volatile acidity.

Rajeshkumar *et al.* (1987) stated the chemical composition of lassi type cultured beverage from cheese whey. According to them, the finished product contains 21.20 per cent total solids, 3.05 per cent protein, 17.85 per cent carbohydrates and 0.30 per cent ash.

Gagrani *et al.* (1987) discussed the Physico-chemical and nutritional qualities of fruit flavoured whey beverage from orange, pineapple, guava, and mango. According to them, orange flavoured beverage contains 20 per cent total solids, 4.30 per cent lactose, 14.06 per cent sucrose, 0.22 per cent proteins, 0.72 per cent ash, 0.5 per cent acidity and 4.21 pH. Pineapple flavoured beverage contains 20 % total solids, 4.18 per cent lactose, 14.16 per cent sucrose, 0.26 per cent proteins, 0.78 per cent ash, 0.5 per cent acidity and 4.22 pH. Guava sucrose, 0.3 per cent proteins, 0.86 per cent ash, 0.5 per cent acidity, 4.29 pH. Mango flavoured beverage contains 20 per cent total solid, 4.20 per cent lactose, 14.15 per cent sucrose, 0.28 per cent proteins, 0.79 per cent ash, 0.5 per cent acidity and 4.23 pH. The colours, per cent sedimentation, per cent turbidity and viscosity (cp) values for orange flavoured beverage were 0.1 R + 0.3 Y, 1.5, 97 and 3.5 respectively; for pineapple flavoured beverage 0.41 Y, 1, 96 and 3.5 respectively. For guava beverage 0.41 Y, 1, 96 and 35 and for mango beverage 0.7 R + 0.45 Y, 2.5, 98 and 4.25 respectively.

Jelen *et al.* (1987) analyzed six commercially whey based beverage from Europe. The total solid, protein, total sugars and pH were in the range of 1.88 – 15.87 per cent, 0.06 – 0.71 per cent, 1.0 – 13.5 per cent and 3.0 – 3.95 per cent respectively.

Reddy *et al.* (1987) studied the physico-chemical composition of whey beverage prepared from lemon juice. According to them, the beverage was of 16 per cent total solids, 0.40 per cent acidity, 5.20 pH.

Krishnaiah *et al.* (1989) analyzed the Physico-chemical composition of beverage prepared from acid whey. They stated that the beverage contains 4.4 pH, 0.413 per cent acidity, 1.26 cp viscosity and 1.078 specific gravity.

Pagote (1993) reported the Physico chemical composition of directly acidified milk based soft drink. According to them the directly acidified milk based soft drink has 22 per cent total solid, 1.7 per cent fat, 1.8 per cent proteins and 3.75 pH.

Singh *et al.* (1994) studied the Physico-chemical composition of whey based beverages. According to them, the beverage contains 13.3 – 16 per cent total solid, 0.32 – 0.38 per cent fat and 0.51 – 0.61 per cent protein.

Khamrui and Rajorhia (1998) reported the Physico-chemical composition of RTS whey based kinnow juice beverage. According to them, the final beverage has 14.25 per cent total solid, 13.09 per cent protein, 0.4 per cent lactose, 2.15 per cent cloudiness, 85 per cent transmission, 7.2 cp viscosity and 4.25 pH.

Wazir Singh *et al.* (1999) determined the Physico chemical composition of soft beverage from paneer whey and guava. The finished product contained 12.33 per cent total solid, 11.68 per cent carbohydrates, 0.34 per cent protein and 0.31 per cent ash.

Kaur *et al.* (2000) studied the physico chemical composition of whey based carrot juice beverage. According to them, the final beverage contained 0.2 per cent acidity, 5.30°Bx, 5.06 pH, 1.012 and specific gravity 26.9 (sec) viscosity.

Sikder *et al.* (2001) studied the physico chemical composition of whey based mango beverage. According to them, the final beverage contained 17.48 per cent total solids, 0.08 per cent fat, 0.2 per cent protein, 6.21 per cent lactose, 8.63 per cent sucrose, 0.49 per cent ash, 0.40 per cent acidity, 1.520 cp viscosity and 1.072 specific gravity.

Shaikh *et al.* (2001) studied the physico chemical composition of fermented carbonated whey beverage. According to them, it contains 83.9 per cent moisture, 0.2 per cent fat, 0.2 per cent protein, 2.98 per cent lactose, 12 per cent sucrose, 0.6 per cent ash, 16.0 per cent total solids, 0.2 per cent acidity and 3.2 pH. It was of 7.91 cp viscosity, 13.1 per cent transmission, 1.50 per cent sedimentation, 0.82 O.D. and 4.2 R + 32.0 Y colour.

Sarvana Kumar and Manimegalai (2002a) reported the physico chemical composition of soymilk whey blended papaya RTS. According to them, the beverage contains 0.3 per cent acidity, 5.34 per cent reducing sugar, 5.60 mg/100 g vitamin C and 14.60 per cent total sugars.

Sarvana Kumar and Manimegalai (2002b) analyzed the Physico-chemical composition of whey based jack fruit RTS beverage. According to them, the final beverage has 17°Bx total soluble solid, 0.20 per cent acidity, 5.2 pH, 5.02 per cent total sugars and 20 per cent reducing sugars.

Sarvana Kumar and Manimegalai (2003) studied the whey based pineapple juice RTS beverage. According to them, the final beverage contained 0.30 per cent acidity, 15°Bx, 5.85 per cent of reducing sugars, 14 per cent total sugars and 3.50 mg/100 g of vitamin C.

Suresha and Jayaprakasha (2003) reported the Physico-chemical properties of lactose hydrolyzed and unhydrolyzed whey permeate beverage. According to them pineapple flavoured beverage contained 12.80 per cent total solids, 0.28 per cent protein, 5.01 per cent lactose, 12.80 per cent sugar, 0.1 per cent acidity, 0.56 per cent ash, 4.48 pH, 1.057 specific gravity and 1.255 cp viscosity. The mango flavoured beverage contained 13.91 per cent total solid, 0.31 per cent proteins, 5 per cent lactose, 12.95 per cent sugar, 0.1 per cent acidity, 0.55 per cent ash, 4.49 pH, 1.058 specific gravity and 1.252 cp viscosity. The orange flavoured beverage contained 13.83 per cent total solid, 0.29 per cent protein, 4.99 per cent lactose, 12.90 per cent sugar, 0.33 per cent acidity, 0.54 per cent ash, 4.52 pH, 1.057 specific gravity and 1.252 cp viscosity.

Shukla et al. (2004) studied the physico chemical composition of the fruit beverages using paneer whey. According to them,mango flavoured beverage contained 662.50 cp viscosity, 16.73 total solids, 0.60 acidity, 0.27 per cent protein and 4.47 pH. Apple flavoured beverage contained 22.50 cp viscosity, 15.22 total solids, 0.54 acidity, 0.34 per cent protein and 4.98 pH. Guava flavoured beverage contained 105.00 cp viscosity, 15.73 total solids, 0.58 acidity, 0.42 per cent protein and 4.76 pH. Banana flavoured beverage contained 130.0 cp viscosity, 18.78 total solids, 0.54 acidity, 0.55 per cent protein and 5.18 pH. Litchi flavoured beverage contained 130.0 cp viscosity, 15.78 total solids, 0.08 per cent acidity, 0.31 per cent protein and 5.76 pH.

MATERIALS AND METHODS

The present investigation was carried out in the Department of Animal Products Technology, College of Agricultural Technology, Marathwada Agricultural University, Parbhani (Maharashtra) during the year 2004-2005. The details of materials used and methods adopted during the present investigation are described in this chapter.

3.1 Materials

3.1.1 Milk

Fresh buffalo milk was obtained from the Department of Animal Husbandry and Dairying, College of Agriculture, Marathwada Agricultural University, Parbhani.

3.1.2 Starter culture

Freeze dried starter culture of *Streptococcus thermophillus* and *Lactobacillus bulgaricus* were obtained from National Dairy Research Institute, Karnal.

3.1.3. Fruit juices

Fresh mango, pineapple and orange juices were obtained from the local market.

3.1.4 Chemicals

Most of the chemicals used in the present investigation were of analytical grade.

3.1.5 Ultrafiltration unit

MILLIPORE make 'Stirred Cell Model -8200' with cell capacity of 200 ml and maximum obtaining pressure of 75 psi. was used. (Figure 2).

3.1.6 Carbonation unit

Laboratory scale carbonation unit was used for the carbonation of beverages.

3.1.7 Preparation of acidic whey (shrikhand whey)

A good quality, fresh buffalo milk was standardized to 6.0 per cent fat by following Pearson's square method. It was heated to about 85ºC for 10 minutes and then cooled to 28-30ºC. It was inoculated with 2.0 percent starter culture and was incubated at 30ºC for 15-16 hours (overnight). When the curd had set firmly (acidity 0.7-0.8 per cent), it was broken and

24

placed in a muslin cloth and hung on a peg for 8-10 hours to remove the whey. During this period, the curd gently squeezed to facilitate whey drainage. Then whey was strained through muslin cloth. The obtained whey was clear and greenish yellow in colour.

3.1.8 Clarification of acidic whey (shrikhand whey)

Initially, the fresh acidic whey (shrikhand whey) was passed through the cream separator for twice at high speed to remove the residual fat. The defatted whey was adjusted to acidity of 0.8 per cent by using citric acid. It was then deproteinized by heating at 98°C for 15 minutes followed by cooling to room temperature. It was kept undisturbed for 5-6 hours to allow the precipitated proteins to settle down. The whey was passed through the cotton pad and double folded muslin cloth (Cheese layer) to get the suspended free product. The supernatant was vaccum filtered through Whatman filter paper No. 41 and cotton pad with muslin cloth to get clear, fat free, greenish yellow coloured whey.

3.1.9 Preparation of prefiltered acidic whey (shrikhand whey)

The clarified acidic whey (shrikhand whey) was prefiltered by passing through 'Microfiber Glassfilter and Mixed Esters of Cellulose' by applying vaccum. The shrikhand whey obtained was clear and free from all the suspended particles.

3.1.10 Preparation of ultrafiltered acidic whey (shrikhand whey)

The prefiltered acidic whey (shrikhand whey) was subjected to ultrafiltration by using 'Polyether Sulphone Biomax Ultrafiltered Disc' having 'Nominal Molecular Weight Limit' (NWML) of 300 KD and diameter of 63.5 mm. in Millipore ultrafiltration unit. The acidic whey (shrikhand whey) obtained was sparkling and clear in appearance.

3.1.11 Fruit juices

Fresh mango, pineapple and orange juices were prefiltered by passing through 'Microfiber Glass Filter and Mixed Esters of Cellulose' by applying vaccum. These were also ultrafiltered by using 'Polyether Sulphone Biomax Ultrafiltered Disc' having 'Nominal Molecular Weight Limit' (NMWL) of 300 KD and diameter of 63.5 mm in Millipore Stirred Cell Ultrafiltration Unit.

3.2 Preparation of whey beverage base

Unclarified acidic whey (shrikhand whey), prefiltered acidic whey (shrikhand whey) and ultrafiltered acidic whey (shrikhand whey) were used for preparation of beverage base. The acidity level in all types of whey was kept constant at 0.8 per cent. To this, sugar was added at the rate of 8, 10, 12 and 14 per cent level in the form of sugar syrup prepared in whey itself. Then, it was pasteurized and cooled to below 50°F temperature. Thus, the prepared whey base samples were subjected to organoleptic evaluation. The whey beverage bases adjudged as a best were used for preparation of fruit flavoured beverages.

3.3 Preparation of fruit flavoured whey beverage

The beverage base adjudged as a best in each category of acidic whey were added with mango, orange and pineapple juices at 18, 20, 22 and 24 per cent level of concentration. For prefiltered whey, prefiltered fruit juices were used and for ultrafiltered whey, ultrafiltered fruit juices were used. Thus prepared beverage was subjected to organoleptic evaluation. The beverages adjudged as best were further studied with respect to Physico-chemical analysis, nutritional aspects and carbonation level.

3.3.1 Flow sheet for fruit flavoured whey beverage

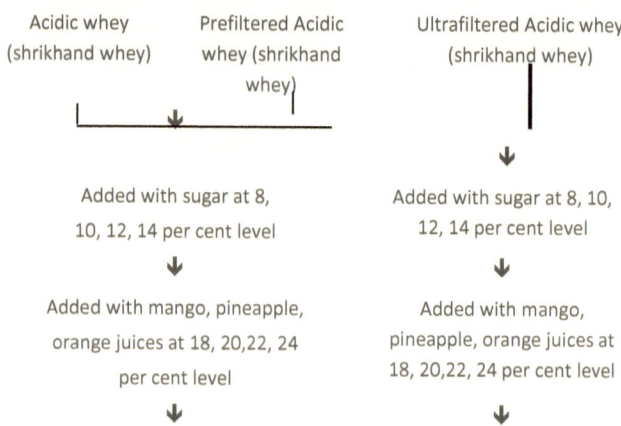

26

Pasteurized at 85°C for 5 minutes	Pasteurized at 85°C for 5 minutes
⬇	⬇
Cooled to room temperature	Cooled to room temperature
⬇	⬇
Bottle filled	Bottle filled
⬇	⬇
Chilled at 4-5°C	Chilled at 4-5°C
⬇	⬇
Carbonated	Carbonated
⬇	⬇
Stored at 7±1°C	Stored at 7±1°C

3.4 Carbonation of whey beverage

The beverages adjudged as best with respect to sugar and fruit juice level were carbonated at three different pressures of 25, 30 and 35 psi at 4±1°C temperature. The beverage was stored at refrigeration temperature till it was subjected to organoleptic evaluation.

3.5 Organoleptic evaluation of whey beverage base and whey beverage

The beverage base and beverages were subjected to organoleptic evaluation by selected judges. It was judged for appearance, colour, taste, aroma, consistency, mouth feel and overall acceptability. The score cards were provided comprising 9-point hedonic scale developed by Quarter Master Foods And Containers Institute, USA (Gupta, 1976).

Sensory Evaluation Score Card

Date:

Name of evaluator:

Trial No.   :

Sample	Appearance	Colour	Taste	Aroma	Consistency	Mouth-feel	Overall acceptability

Score card: 9 – Like extremely

8 – Like very much

7 – Like moderately

6 – Like slightly

5 – Neither like nor dislike

4 – Dislike slightly

3 – Dislike moderately

2 – Dislike very much

1 – Dislike extremely

Comments, if any –

Signature of evaluator

These score cards were filled in, every time by the judges. The score given by the judges were statistically analysed.

3.6 Analytical method

3.6.1 Physical properties

3.6.1.1 Determination of colour

The colour of acidic whey sample was measured by using Lovibond Tintometer and specified by the values of red, yellow and blue slides required for comparison.

3.6.1.2 Determination of specific gravity

The specific gravity of acidic whey sample was estimated at 20°C by using a standard specific gravity bottle of 50 ml capacity taking distilled water as standard liquid.

$$\text{Specific gravity} = \frac{\text{Density of acidic whey sample}}{\text{Density of water}}$$

3.6.1.3 Determination of viscosity

It was performed by using 'Haake's Roto Viscometer,' RV-20 Model (Germany) calibrated at 32°C by using NV system at (S^{-1}) shear rate. The dial reading was recorded as % T and calculations were made by using following equation.

T	=	% T x A
Where, T	=	Shearing stress (Pa)
%T	=	Shear stress value of the display
A	=	Shear stress factor
D	=	% D x M
Where, D	=	Shearing rate (S^{-1})
%D	=	Shear rate set
M	=	Shear rate factor

The spindle constant A (1.78) and M (27.0) for particular spindle NV were obtained from the Manual of the instrument supplied by manufacturing company.

The calculations for apparent viscosity were made by using following formula.

$$N \quad = \quad T/D. \text{ (Pas)}$$
$$N \quad = \quad T/D \times 1000 \text{ (cp)}$$

3.6.1.4 Determination of sedimentation

It was estimated by centrifugation method. 10 ml of acidic whey sample was transferred to centrifuge tube and centrifuged at 10,000 rpm for 20 min. The supernatant liquid was decanted and measured. The difference between initial volume and final volume gave sediments present in 10 ml and expressed as per cent sedimentation.

3.6.1.5 Determination of turbidity

The turbidity of acidic whey sample was measured by using turbidity meter. Initially, the turbidity meter was standardized by using distilled water. The acidic whey sample was placed into cuvette of turbidity meter and light was passed through sample. The needle on the dial of turbidity meter directly indicated the turbidity of the whey sample in per cent.

3.6.2 Proximate composition

It was estimated for acidic whey (shrikhand whey), prefiltered acidic whey (shrikhand whey) and ultrafiltered acidic whey (shrikhand whey) and for fruit flavoured whey beverage.

3.6.2.1 Estimation of moisture and total solids

The clean dry empty dish was weighed accurately. 5 ml of acidic whey sample was pipetted in to dish and weighed quickly. Dish was placed on boiling water bath. After 30 minutes, dish was removed wiped the bottom and transferred to well ventilated oven at 98-100°C temperature. After 2-3 hours the dish was immediately transferred to desiccator allowed to cool for about 30 min. and weighed. The dish was returned to oven and heated for 1 more hour. It was removed from desiccators, cooled and weighed as before repeated of necessary until the loss in weight between successive weights didn't exceed by 0.5 mg (I.S.I., SP : 18, Part – XI, 1981). The per cent moisture and total solids were calculated by following formula.

$$\% \text{ Moisture} = \frac{W_1}{W_2} \times 100$$

. W_1 =   Weight loss in 'g' of acidic whey sample

W_2 =   Weight in 'g' of acidic whey sample

$$\% \text{ Total solids} = \frac{100 \times W_1}{W_2}$$

. W_1 =   Weight loss in 'g' of the residue after drying

W_2 =   Weight in 'g' of acidic whey sample

3.6.2.2 Estimation of fat

Fat content was determined as per the procedure described in I.S.I. (SP : 18) Part – XI, 1981.

3.6.2.3 Estimation of protein

It was estimated by Micro-kjeldhal method using 0.5 g of acidic whey sample; by digesting the same with concentrated H_2SO_4 containing catalyst mixture for 3-4 hrs. at 100°C. It was then distilled with 40 per cent NaOH and liberated ammonia was trapped in 4 per cent boric acid using mixed indicator (methyl red : bromocressol green = 1:3). The per cent nitrogen was calculated and protein content was estimated by multiplying with factor 6.38 (A.O.A.C., 1990).

$$\% \text{ Nitrogen} = \frac{\text{Sample - blank} \times \text{Normality} \times \text{Dilution} \times 14.00}{\text{Weight of sample (mg)}} \times 100$$
$$\text{titre} \quad \text{titre} \quad \text{HCl} \quad \text{factor}$$

$\% \text{ Protein}$ = $\% \text{ N} \times 6.38$

3.6.2.4 Estimation of lactose

Lactose was determined as per the procedure described in 'Practical Agricultural Chemistry' by Kanwar and Chopra (1976). 10 ml of whey sample was taken in 250 ml flask. To this 1-2 ml of 10 per cent acetic acid was added and shaken well. The contents were diluted by adding 90 ml of distilled water. The contents of the flask were heated, cooled and volume was made to 100 ml by distilled water in a measuring cylinder and shaken well. This content were filtered and filtrate was filled in a burette and titrated against 10 ml of benedicts solution + 1 g of sodium carbonate. The titration was done in hot condition and continued till blue colour disappeared. The burette reading was noted and per cent lactose was determined by following formula.

$$\% \text{ Lactose} = \frac{\text{Sample (ml)} \times 0.268}{\text{Burette reading}} \times 100$$

3.6.2.5 Estimation of total ash

It was carried out according to I.S.I. (SP:18) Part – XI, 1981. The acidic whey sample (5g) was weighed in previously weighed silica crucible. The sample was evaporated to dryness on a heater and ignited in a muffle furnace at a temperature not more than 550°C until the ash was completely free from carbon. Then, cooled in a desiccator and weighed accurately and was repeated until two consecutive weights were constant.

$$\% \text{ Ash} = \frac{100 \times W_1}{W_2}$$

W_1 = Weight in 'g' of ash

W_2 = Weight in 'g' of acidic whey sample

3.6.2.6 Estimation of titratable acidity

It was determined as per the method cited in I.S.I. (SP:18) Part-XI, 1981. The acidic whey sample was mixed thoroughly. Ten ml of this was pipetted and poured in to two porcelian dish. Equal volume of freshly boiled, cooled, distilled water was added. To one dish, 1 ml of phenolphthalein solution was added and to other rosaniline acetate was added. The contents of dish containing phenolphthalein was titrated against 0.1 N NaOH. This was added drop by drop from the burette until the colour matches to pink tint of rosaniline acetate solution. It was expressed as lactic acid per 100 ml of whey.

$$\% \text{ Titratable acidity} = \frac{9 \times V_1 \times N}{V_2}$$

V_1 = Volume of NaOH (ml)

N = Normality of NaOH

V_2 = Volume of sample (ml) taken for test

3.7 Estimation of Physico-chemical quality of acidic whey (shrikhand whey) beverage

It was determined as per the methods cited in **3.6.1** and **3.6.2.**

3.8 Nutritional aspects of acidic whey (shrikhand whey) beverage

3.8.1 Determination of calcium

Calcium content of acidic whey beverage was determined by using Flame Photometer. The sample was digested with tri-acids i.e nitric acid, per chloric acid and sulphuric acid (3:2:1). The calcium content was determined from the tri acid digested sample by using Flame Photometer as per the method described by Tondon (1993).

3.8.2 Estimation of phosphorus

The phosphorus content of the acidic whey beverage was determined as per the method cited in A.O.A.C. (1990). The obtained ash of the beverage sample was added with 0.5 – 1ml of glass distilled water and 5 ml of concentrated HCl. It was then evaporated on water bath. It was again added with 4 ml of concentrated HCl and 5 ml of distilled water, warmed and filtered through Whatman No. 4. The final volume was made to 100 ml using distilled water. 0.5 ml of this solution was taken in 25 ml volumetric flask to which each 1 ml of ammonium molybdate-sulphuric acid reagent, hydroquinone (0.5%) sodium sulphite (20%) were added in order with subsequent shaking after each addition. Finally, volume was made to 25 ml and optical density was measured at 660 nm within 30 min. Similarly, a blank was also run as such. The phosphorus content was calculated from standard curve prepared with standard phosphate solution (0.4394 g/lit).

1 ml of standard solution = 0.01 mg P

3.8.3 Determination of ascorbic acid

The ascorbic acid content of whey beverage was determined as per the procedure cited in Raganna (1986). 5 ml of working standard solution of 1 μg ascorbic acid was taken in 100 ml conical flask. To this 10 ml of 4% oxalic acid was added and titrated against 2-6 dichlorophenol indophenol dye till the end point of pink colour persisted for few minutes (V_1ml). The beverage sample was taken instead of standard solution and same procedure was followed. (V_2 ml)

33

$$\text{Ascorbic acid (mg/100g)} = \frac{0.5}{V_1\ ml} \times \frac{V_2\ ml}{15} \times \frac{100}{Wt.\ of\ sample} \times 100$$

3.8.4 Determination of vitamin A

Vitamin A content of acidic whey beverage was determined with HPLC (High Performance Liquid Chromatography) method as described by Suzanne Nielsen (1994).

The beverage sample (10 g) was taken and to this added 2 ml of 50 per cent KOH and 50 ml of ethanol and kept in water bath at 80°C to saponify. Cooled in ice bath and extracted in petroleum ether : diethyl ether (50 : 50). Washed the organic layer with water till alkali free. Evaporated residues at 40°C under nitrogen and dissolved in methanol : water (95:5) mobile phase and was detected in HPLC on UV detector at 325 nm.

$$\text{Vit A}\ (\mu g/g) = C \times \frac{DV}{WT}$$

C   =   Vitamin A concentration resulting from sample and standard peak height

DV   =   Final dilution volume of sample

WT   =   Sample weight (g)

3.9 Statistical analysis

The data obtained for various characteristics were complied, tabulated and statistically analyzed by completely randomized design as per the method given by Panse and Sukhatme (1985) in order to draw a meaningful conclusion.

RESULTS AND DISCUSSION

In the present study, attempts have been made to standardize the process for the manufacture of carbonated fruit flavoured beverage by using acidic whey (Shrikhand whey), the major by-product of dairy industry. The beverage prepared was assessed for its sensory quality. The selected beverages were analyzed for their Physico-chemical properties and nutritional aspects. The results recorded during the present investigation are presented under suitable headings. The results are also discussed in the view of relevant scientific literature available in the country and else where.

4.1   Physico-chemical composition of unclarified, prefiltered and ultrafiltered acidic whey (shrikhand whey).

4.2   Physico-chemical composition of unclarified, prefiltered and ultrafiltered fruit juices.

4.3   Standardization of acidic whey (shrikhand whey) beverage base.

4.4   Effect of levels of various fruit juices on sensory quality of acidic whey (shrikhand whey) beverage.

4.5   Physico-chemical composition of acidic whey (shrikhand whey) beverage.

4.6   Nutritional quality of acidic whey (shrikhand whey) beverage.

4.7   Carbonation of acidic whey (shrikhand whey) beverage.

4.1   Acidic whey (shrikhand whey)

4.1.1. Physical properties

The physical properties viz. colour, specific gravity, viscosity, sedimentation and turbidity of different types of acidic whey (shrikhand whey) were determined and are tabulated in Table 1.

Table 1. Physical properties of unclarified, prefiltered and ultrafiltered acidic whey (shrikhand whey)

Physical properties	Unclarified	Clarified	
	acidic whey (shrikhand whey)	Prefiltered acidic whey (shrikhand whey)	Ultrafiltered acidic whey (shrikhand whey)

Colour	0.3 R + 1.3 Y	0.1 R + 1 Y	0.1 R + 1 Y
Specific gravity	1.030	1.017	1.015
Viscosity (cp)	3.48	3.10	1.24
Sedimentation (%)	7.52	3.20	1.52
Turbidity (%)	100.00	87.00	4.00

(Figure in parenthesis indicates mean value)

The results of the Table 1 indicates that the colour of unclarified acidic whey (shrikhand whey) was recorded as 0.3 R + 1.3 Y, where as there is a change in the colour of prefiltered and ultrafiltered acidic whey (shrikhand whey) i.e. there is decrease in the red value and yellow value as compared to unclarified acidic whey (shrikhand whey).

It can be further seen from Table 1 that the specific gravity of unclarified acidic whey (shrikhand whey) was recorded as 1.030. There is decrease in the values of the specific gravity of the prefiltered and ultrafiltered acidic whey (shrikhand whey). This is a indicative of removal of large size particles during the process of filtration which has reduced the value of specific gravity.

The viscosity of acidic whey (shrikhand whey) was determined by Hakke's Rotoviscometer and is expressed in centipoises. The viscosity of the unclarified acidic whey (shrikhand whey) was much higher than that of prefiltered and ultrafiltered acidic whey (shrikhand whey). This is due to presence of fat and higher amount of protein.

The sedimentation values of the unclarified acidic whey (shrikhand whey) was of 7.52 per cent where as that of prefiltered and ultrafiltered acidic whey (shrikhand whey), the sedimentation values were 3.20 and 1.52 per cent respectively. The lowest sedimentation value was observed in ultrafiltered acidic whey (shrikhand whey) which may be due to negligible amount of constituents present in it, which are responsible for lowering sedimentation value.

It can be further revealed from Table 1 that the turbidity per cent was in the range of 100, 87 and 4 per cent in the samples under study. The lowest turbidity value in case of ultrafiltered acidic whey (shrikhand whey) is due to minimum presence of constituents having less adverse effect on the transmission of light resulting in to lower values of turbidity.

The specific gravity and viscosity values for ultrafiltered acidic whey (shrikhand whey) reported by Suresha and Jayaprakasha (2003) are in accordance with the findings of the present study.

4.1.2   Proximate composition

The proximate composition of acidic whey (shrikhand whey), prefiltered acidic whey (shrikhand whey) and ultrafiltered acidic whey (shrikhand whey) used in this study is depicted in Table 2.

Table 2 Proximate composition of unclarified, prefiltered and ultrafiltered acidic whey (shrikhand whey)

Chemical constituents (%)	Unclarified	Clarified	
	acidic whey (shrikhand whey)	Prefiltered acidic whey (shrikhand whey)	Ultrafiltered acidic whey (shrikhand whey)
Moisture	93.43	94.04	94.38
Total solids	6.57	5.96	5.62
Fat	0.20	---	---
Protein	0.65	0.36	0.29
Lactose	4.90	4.85	4.78
Ash	0.76	0.52	0.50
Acidity	0.70	0.81	0.81
pH	4.50	4.36	4.36

(Figure in parenthesis indicates mean value)

From the data presented in Table 2, it reveals that the total solids content was found to be 6.57 per cent in unclarified acidic whey (shrikhand whey), which get reduced in prefiltered and ultrafiltered acidic whey (shrikhand whey). The similar trend was observed with reference to protein, lactose and ash. However, the protein content has been reduced to a greater extent i.e. from 0.65 per cent in unclarified acidic whey (shrikhand whey) to 0.29 per cent in ultrafiltered shrikhand whey, which is

37

mainly due to the deporteinization carried out during for ultrafiltration process.

It can be further observed from Table. 2 that fat content of unclarified acidic whey (shrikhand whey) was only 0.20 per cent but there was no detectable fat in prefiltered and ultrafiltered shrikhand whey. This is due to reseperation of acidic whey (shrikhand whey) and process of filtration also helped in removal all of the fat to a greater extent.

It can be also observed from Table 2. that acidity of unclarified and clarified acidic whey (shrikhand whey) was in the range of 0.70 per cent to 0.81 per cent with the corresponding pH in the range of 4.50 – 4.36. The higher acidity in clarified acidic whey (shrikhand whey) is due to more conversion of lactose in to lactic acid as the clarified acidic whey (shrikhand whey) was stored for certain period in undistributed state in order to allow the proteins to settle at the bottom.

Kulkarni (1987) reported that acidic whey (shrikhand whey) contains 94.07 per cent moisture, 5.93 per cent total solids, 0.18 per cent fat, 0.65 per cent protein, 4.59 per cent lactose, 0.55 per cent ash, and 0.9 per cent acidity. Jadhav (1991) also reported that acidic whey (shrikhand whey) contains 91.95 per cent moisture, 8.02 per cent total solids, 0.35 per cent fat, 0.52 per cent ash, 0.78 per cent acidity and 4.66 pH. Khamrui and Rajorhia (1998) revealed that acidic whey (shrikhand whey) contains 93.04 moisture 6.96 pr cent total solids, 0.43 per cent fat, 0.38 per cent protein and 0.94 per cent acidity.

The results of present study are in good confirmation with those mentioned above.

4.2 Fruit juices

4.2.1 Physical properties

The fruit juices adjudged as the best for the preparation of the final beverage were studied for their physical properties viz. colour specific gravity, viscosity, sedimentation and turbidity are tabulated in Table 3.

Table 3.   Physical properties of unclarified mango juice, prefiltered orange juice and ultrafiltered pineapple juice

Physical properties	Unclarified Mango juice	Prefiltered Orange juice	Ultrafiltered Pineapple juice
Colour	1R + 10 Y	1.2 R + 14 Y	1Y + 0.1 B

Specific gravity	1.060	1.052	1.050
Viscosity (cp)	1.520	1.350	1.252
Sedimentation (%)	12.00	8.00	4.00
Turbidity (%)	100.00	85.00	5.00

(Figure in parenthesis indicates mean value)

The results of the Table 3 indicate that the colour of unclarified mango juice was 1 R + 10Y. It also revealed that the specific gravity of the unclarified mango juice was found to be 1.060. The viscosity of the unclarified mango juice was 1.520 cp. The sedimentation value for unclarified mango juice was 12 per cent. The turbidity was found to be 100 per cent for unclarified mango juice.

The results of the Table 3 indicates that the colour of prefiltered orange juice was 1.2 R + 14Y. It also revealed that the specific gravity of the prefiltered orange juice was found to be 1.052. The viscosity of the prefiltered orange juice was 1.350 cp. The sedimentation value for prefiltered orange juice was 8 per cent. The turbidity was found to be 85 per cent for prefiltered orange juice.

The results of the Table 3 indicates that the colour of ultrafiltered pineapple juice was 1 Y + 0.1 B. It also revealed that the specific gravity of the ultrafiltered pineapple juice was found to be 1.050. The viscosity of the ultrafiltered pineapple juice was 1.252 cp. The sedimentation value for ultrafiltered pineapple juice was 4 per cent. The turbidity was found to be 5 per cent for ultrafiltered pineapple juice. The lowest turbidity value in case of ultrafiltered pineapple juice is due to minimum presence of constituents having less adverse effect on transmission of light.

The specific gravity and viscosity values for mango juice reported by Shukla *et al.* (2004) are in accordance with the findings of present study.

4.2.2 Proximate composition of fruit juices

The proximate composition of unclarified mango juice, prefiltered orange juice and ultrafiltered pineapple juice used in this study is depicted in Table 4.

Table 4. Proximate composition of unclarified mango juice, prefiltered orange juice and ultrafiltered pineapple juice

Chemical constituents (%)	Unclarified Mango juice	Prefiltered Orange juice	Ultrafiltered Pineapple juice
Moisture	84.50	94.04	94.38
Total solids	15.50	5.96	5.62
Fat	0.1	---	---
Protein	0.2	0.1	0.1
Ash	0.60	0.40	0.29
Acidity	0.31	0.30	0.30
pH	4.23	4.23	4.23

(Figure in parenthesis indicates mean value)

From the data presented in Table 4, it revels that the total solids content was found to be 15.5 per cent in unclarified mango juice.

The fat content of unclarified mango juice was found to be 0.1 per cent. The protein content for unclarified mango juice was found to be 0.2 per cent. The ash content was 0.6 per cent. The acidity for unclarified mango juice was 0.31 per cent with corresponding pH 4.23.

From the data presented in Table 4, it reveals that the total solids content was found to be 5.96 per cent in prefiltered orange juice,

The protein content for prefiltered orange juice was found to be 0.1 per cent. The ash content was 0.4 per cent. The acidity for prefiltered orange juice was 0.30 per cent with corresponding pH 4.23. There was no detectable fat in prefiltered orange juice. It is due to reseperation of orange juice and process of filtration also helped in removal all of the fat.

From the data presented in Table 4, it revels that the total solids content was found to be 5.62 per cent in ultrafiltered pineapple juice. The protein content for ultrafiltered pineapple juice was found to be 0.1 per cent. The ash content was 0.29 per cent. The acidity for unclarified mango juice was 0.30 per cent with corresponding pH 4.23. There was no detectable fat in prefiltered orange juice. It is due to reseperation of orange juice and process of filtration also helped in removal all of the fat.

There was no detectable protein in ultrafiltered pineapple juice due to deproteinization carried out during ultrafiltration process.

The results of present study are in good confirmation with the results reported by Shukla *et al.* (2004).

4.3 Standardization of acidic whey (shrikhand whey) beverage base

Unclarified, prefiltered and ultrafiltered acidic whey (shrikhand whey) were used to standardize the beverage base with respect to levels of sugar on the basis of sensory quality. The related data are shown in Table 5 with respect to appearance, colour, taste, aroma, consistency, mouth feel and overall acceptability.

The higher value for overall acceptability were recorded as 8.10, 8.30 and 8.40 in unclarified, prefiltered and ultrafiltered acidic whey (shrikhand whey) beverage base respectively for 12 per cent sugar level.

The higher overall acceptability score in case of unclarified acidic whey (shrikhand whey) beverage base was due to improvement in taste and mouth feel, whereas in case of prefiltered and ultrafiltered beverage base, the highest score was awarded to all parameters. The effect of sugar level on appearance and colour parameters was found to be non significant in all three types of beverage bases.

It can be further stated from Table 5 that the highest score for overall acceptability was 8.40 for ultrafiltered acidic whey (shrikhand whey) beverage base for 12 per cent sugar level. This indicates that it was 'liked very much' by the judges. The beverage base prepared from unclarified and prefiltered acidic whey (shrikhand whey) beverage base was also rated as a 'liked very much' (8.10 and 8.30 respectively).

From the data of Table 5, it can be revealed that the acidic whey (shrikhand whey) beverage base prepared from 12 per cent sugar level is adjudged as the best for all types of shrikhand whey. The increase in sugar level over 12 per cent adversely affected the sensory quality of beverage base to a great extent in all the three types of acidic whey (shrikhand whey). The results were also found to be statistically significant for all parameters of sensory quality.

Thus, the results of the present study are in accordance with the results of Shaikh *et al.* (2001).

41

Table 5. Effect of levels of sugars on sensory quality of beverage base

Sugar (%)	Appearance (%)	Colour	Taste	Aroma	Consistency	Mouth feel	Overall accept-ability
8	7.60	7.30	6.10	7.00	7.50	7.50	6.60
10	7.60	7.50	7.10	7.00	7.50	7.50	7.30
12	**7.60**	**7.60**	**8.00**	**7.00**	**7.80**	**8.30**	**8.10**
14	7.80	7.80	7.80	7.30	7.60	7.60	7.80
SE	NS	NS	0.22	0.02	0.21	0.02	0.02
CD at 5%	---	---	0.56	0.06	0.61	0.06	0.08
Prefiltered acidic whey (shrikhand whey)							
8	7.20	7.60	6.80	6.90	6.80	7.10	6.80
10	7.45	7.50	6.70	7.10	7.10	7.10	6.90
12	**8.25**	**7.60**	**8.40**	**8.30**	**8.40**	**8.70**	**8.30**
14	8.30	7.50	7.40	8.10	8.20	7.50	7.30
SE	NS	NS	0.29	0.32	0.29	0.27	0.23

CD at 5%	---	---	0.88	0.96	0.88	0.83	0.71
Ultrafiltered acidic whey (shrikhand whey)							
8	7.40	7.50	6.40	6.20	6.90	6.60	6.50
10	7.70	7.90	7.00	7.40	7.50	7.30	7.60
12	7.70	8.10	8.30	8.40	8.40	8.60	8.40
14	7.50	7.70	7.50	7.90	7.40	7.20	7.50
SE	NS	NS	0.3	0.3	0.28	0.34	0.28
CD at 5%	---	---	0.9	0.9	0.85	1.03	0.84

4.4   Effect of levels of various fruit juices on sensory quality of acidic whey (shrikhand whey) beverage

The acidic whey (shrikhand whey) beverage adjudged as a best for 12 per cent sugar level in all three types of acidic whey (shrikhand whey) beverage combined with different fruit juices at various levels of concentration to know the most acceptable fruit flavoured acidic whey (shrikhand whey) beverage on the basis of sensory quality.

4.4.1   Acidic whey (shrikhand whey) beverage from mango juice

The effect of mango juice and its level of concentration on sensory quality of unclarified, prefiltered and ultrafiltered acidic whey (shrikhand whey) beverage is expressed in Table 6.

From Table 6 it can be observed that the highest score of 8.30 and 8.30 was recorded with respect to appearance and colour of unclarified acidic whey (shrikhand whey) beverage prepared from mango juice at 22 per cent level of concentration.

Table 6 further indicates that highest consistency score of 8.06 was recorded for unclarified acidic whey (shrikhand whey) beverage prepared from 22 per cent mango juice concentration.

It can be also noted that the highest overall acceptability scores were recorded for beverages prepared from unclarified, prefiltered and ultrafiltered acidic whey (shrikhand whey) at 22 per cent mango juice concentration i.e. 8.23, 7.00 and 6.75 respectively.

From the data of Table 6 it can be deduced that 22 per cent level of concentration of mango juice with unclarified acidic whey (shrikhand whey) is adjudged as best for acidic whey (shrikhand whey) beverage preparation.

The results of the present study are in close confirmation with those reported by Gagrani *et al.* (1987). The results were also found to statistically significant and the level of concentration of fruit juice had significant effect on all parameters of sensory quality irrespective of type of acidic whey (Shrikhand whey) used.

Table 6. Effect of levels of mango juice on sensory quality of acidic whey (shrikhand whey) beverage

Type of acidic whey (shrikhand whey)	Fruit juice (%)	Appearance (%)	Colour	Taste	Aroma	Consistency	Mouth feel	Overall acceptability
Unclarified	18	6.65	5.95	6.00	5.40	5.80	6.00	5.90
Prefiltered	18	5.85	5.75	5.95	5.60	5.90	5.75	5.96
Ultrafiltered	18	6.50	5.75	5.65	5.75	5.80	5.75	5.95
Unclarified	20	6.13	6.33	5.90	5.80	5.93	6.06	5.93
Prefiltered	20	5.90	5.80	6.03	5.73	6.33	5.86	6.03
Ultrafiltered	20	6.53	5.83	5.76	5.86	5.93	5.76	6.03
Unclarified	**22**	**8.30**	**8.30**	**7.96**	**7.93**	**8.06**	**8.26**	**8.23**
Prefiltered	22	6.96	7.0	6.93	6.90	7.00	7.06	7.00
Ultrafiltered	22	6.86	6.5	6.90	6.80	6.96	6.75	6.75
Unclarified	24	6.50	6.20	6.23	6.20	6.63	6.40	6.20
Prefiltered	24	5.93	5.90	6.10	5.90	6.03	6.00	6.10

Ultrafiltered	24	6.60	6.10	6.03	5.96	6.00	5.86	6.20
SE	NS	NS	0.13	0.16	0.14	0.12	0.14	0.81
CD@5%	--	--	0.4	0.47	0.41	0.37	0.41	0.23

(Each value is mean score of six judges in three trials)

4.4.2   Acidic whey (shrikhand whey) beverage from orange juice

The effect of levels of orange juice on the sensory quality of unclarified, prefiltered and ultrafiltered beverage is tabulated in Table 7.

The observations recorded in Table 7 indicate that the appearance and colour score of prefiltered acidic whey (shrikhand whey) at 22 per cent level of concentration of orange juice was highest i.e. 8.06 and 8.30 respectively.

It can be observed from Table 7 that the lowest score for the taste parameter was 6.23 i.e. 'liked moderately' for unclarified acidic whey (shrikhand whey) beverage whereas prefiltered acidic whey (shrikhand whey) and ultrafiltered acidic whey (shrikhand whey) obtained the score of 8.20 and 6.20 respectively.

From Table 7, it can be further noted that aroma score of unclarified, prefiltered and ultrafiltered acidic whey (shrikhand whey) was 6.93, 8.10 and 7.23 respectively at 22 per cent orange juice concentration.

Table 7 revealed that the consistency scores of unclarified, prefiltered and ultrafiltered acidic whey (shrikhand whey) beverage was 6.80, 8.20 and 8.10 respectively. It means the prefiltered and ultrafiltered acidic whey (shrikhand beverage) are 'liked very much' by the judges.

From the data of Table 7 it can be deduced that 22 per cent level of concentration of orange juice with prefiltered acidic whey (shrikhand whey) is adjudged as best for acidic whey (shrikhand whey) beverage preparation.

Table 7. Effect of levels of orange juice on sensory quality of acidic whey (shrikhand whey) beverage

Type of acidic whey (shrikhand whey)	Fruit juice (%)	Appearance (%)	Colour	Taste	Aroma	Consistency	Mouth feel	Overall acceptability
Unclarified	18	6.00	5.95	8.85	6.00	8.85	5.95	5.85
Prefiltered	18	6.00	5.85	6.00	5.95	5.75	6.00	5.95
Ultrafiltered	18	6.06	5.95	5.95	5.95	6.20	5.85	6.25
Unclarified	20	6.00	6.00	6.00	6.15	6.00	6.05	6.06
Prefiltered	20	6.10	5.95	6.10	6.00	5.95	6.35	0.10
Ultrafiltered	20	5.95	6.00	6.03	6.05	6.05	6.45	6.00
Unclarified	22	6.56	6.46	6.23	6.93	6.80	6.46	6.43
Prefiltered	**22**	**8.06**	**8.30**	**8.20**	**8.10**	**8.20**	**7.03**	**7.96**
Ultrafiltered	22	6.20	6.23	6.20	7.23	8.10	6.70	6.80
Unclarified	24	6.36	6.35	6.15	6.25	6.10	6.13	6.16
Prefiltered	24	6.33	6.15	6.40	5.53	7.20	6.36	6.83

Ultrafiltered	24	6.06	6.15	6.20	6.30	6.23	5.60	6.56
SE	--	0.98	0.05	0.75	0.70	0.10	0.75	0.80
CD@5%	--	0.16	0.19	0.22	0.20	0.35	0.22	0.34

(Each value is mean score of six judges in three trials)

From Table 7, it is noted that overall acceptability of prefiltered acidic whey (shrikhand whey) beverage at 22 per cent level of concentration of orange juice was higher i.e. 7.96 and that of unclarified and ultrafiltered beverage was 6.43 and 6.80 respectively. Statistically there was significant effect of concentration on mouth feel.

Shaikh *et al.* (2001) recorded 7.60 scored for overall acceptability of orange flavoured whey beverage.

Gagrani *et al.* (1987) recorded average mean score of 7.06 for orange flavoured whey beverage with 25 per cent level of concentration. The results of the present findings are comparable to those mentioned above.

4.4.3 Acidic whey (shrikhand whey) beverage form pineapple juice

The effect of levels of pineapple juice on the sensory quality of unclarified, prefiltered, ultrafiltered acidic whey (shrikhand whey) beverage is shown in Table 8.

From the results of Table 8, it is clear that the appearance and colour attributes scored was highest for ultrafiltered acidic whey (shrikhand whey) prepared from pineapple juice at 22 per cent level of concentration. The average score for aroma was found to be 7.00, 6.83 and 8.00 for unclarified, prefiltered and ultrafiltered acidic whey (shrikhand whey) beverage respectively.

The highest score of 8.10 i.e. 'liked very much' was obtained to the ultrafiltered acidic whey (shrikhand whey) beverage as compared to unclarified and prefiltered having 6.53 and 6.46 scores respectively.

It can be observed from Table 8 that overall acceptability score for ultrafiltered acidic whey (shrikhand whey) beverage was superior i.e. 8.26 to that of unclarified and prefiltered acidic whey (shrikhand whey) beverage.

Table 8. Effect of levels of pineapple juice on sensory quality of acidic whey (shrikhand whey) beverage

Type of acidic whey (Shrikhand whey)	Fruit juice (%)	Appearance (%)	Colour	Taste	Aroma	Consistency	Mouthfeel	Overall acceptability
Unclarified	18	5.85	5.85	5.85	5.95	5.80	5.95	5.90
Prefiltered	18	5.90	5.75	5.95	5.75	5.95	5.75	5.95
Ultrafiltered	18	6.00	5.95	6.00	5.80	6.10	5.85	6.01
Unclarified	20	5.95	6.10	5.95	6.15	6.00	5.85	6.00
Prefiltered	20	6.00	6.23	6.15	6.03	6.25	5.85	6.06
Ultrafiltered	20	6.35	6.3	6.25	6.13	6.40	6.15	6.33
Unclarified	22	6.26	6.83	6.83	7.00	6.53	6.53	6.43
Prefiltered	22	6.86	7.50	6.70	6.83	6.86	6.46	6.60
Ultrafiltered	**22**	**7.10**	**8.30**	**8.10**	**8.00**	**8.06**	**8.10**	**8.26**
Unclarified	24	6.16	6.43	6.20	6.26	6.43	6.09	6.23
Prefiltered	24	6.46	7.03	6.36	6.56	6.66	6.06	6.36

Ultrafiltered	24	6.70	7.56	6.96	6.23	7.10	7.00	6.53
SE	NS	0.07	0.06	0.07	0.11	0.10	0.11	0.04
CD@5%	--	0.22	0.21	0.22	0.33	0.33	0.34	0.14

(Each value is mean score of six judges in three trials)

Suresha and Jayaprakasha (2003) reported that pineapple flavour whey permeate beverage was found to be most acceptable and had highest mean overall acceptability score of 8.65 i.e. 'liked very much'. The results were also found to be statistically significant for all parameters of sensory quality.

Shaikh *et al.* (2001) observed that overall acceptability score of 8.06 for pineapple flavoured whey beverage and also reported that this beverage was better than orange flavoured beverages.

4.5 Acidic whey (shrikhand whey) beverage

The beverage adjudged as a best for each fruit juice was selected for further studies. The beverage from mango flavoured unclarified acidic whey (shrikhand whey) [A], orange flavoured prefiltered acidic whey (shrikhand whey) [B] and pineapple flavoured ultrafiltered acidic whey (shrikhand whey) [C] were used for further studies with respect to physical, chemical and nutritional aspects.

4.5.1 Physical properties of acidic whey (shrikhand whey) beverage

The physical properties viz. colour, specific gravity, viscosity, sedimentation, turbidity of selected acidic whey (shrikhand whey) beverage were determined and tabulated in Table 9.

Table 9. Physical properties of acidic whey (shrikhand whey) beverage

Physical properties	Unclarified	Clarified	
	A	B	C
Colour	1R + 13Y	1.4R + 14.1Y	1Y + 0.1B
Sp. gravity	1.072	1.060	1.057
Viscosity (cp)	1.520	1.355	1.252
Sedimentation (%)	12.00	8.00	4.00
Turbidity (%)	100.00	92.00	5.00

(Figures in parenthesis indicate mean value)

It can be noted from Table 9 that the colour of unclarified mango flavoured acidic whey (shrikhand whey) beverage was 1R + 13Y, for prefiltered orange flavoured acidic whey (shrikhand whey) beverage was 1.4R + 14.1 and for ultrafiltered pineapple flavoured acidic whey (shrikhand whey) beverage was 1Y + 0.1B. The 'Y' value of unclarified mango flavoured and prefiltered orange flavoured acidic whey (shrikhand whey) beverage was higher than ultrafiltered pineapple flavoured acidic whey (shrikhand whey) beverage. It is due to higher β–carotene content of it.

The specific gravity values (Table 9) were in the range of 1.072 to 1.057 with higher value of unclarified mango flavoured acidic whey (shrikhand whey) beverage and lowest value of ultrafiltered pineapple flavoured acidic whey (shrikhand whey) beverage. This variation is due to the difference in the total solids content of the beverages.

From Table 9 it can be observed that viscosity of unclarified mango flavoured acidic whey (shrikhand whey) beverage was higher than that of prefiltered orange flavoured and ultrafiltered pineapple flavoured acidic whey (shrikhand whey) beverage.

Sedimentation value of unclarified mango flavoured, prefiltered orange flavoured and ultrafiltered pineapple flavoured acidic whey (shrikhand whey) beverage was 12.00, 8.00 and 4.00 per cent respectively. The lowest value was found in ultrafiltered pineapple flavoured acidic whey (shrikhand whey) beverage due to negligible amounts of constituents responsible for sedimentation value.

The per cent turbidity of ultrafiltered pineapple flavoured acidic whey (shrikhand whey) beverage was only 5 per cent. It is very much lower than that of unclarified mango flavoured and prefiltered orange flavoured beverage due to decrease in protein content during ultrafiltration process which increased the clarity of beverage (Table 9).

The specific gravity and viscosity value for ultrafiltered pineapple flavoured acidic whey (shrikhand whey) beverage reported by Suresha and Jayaprakash (2004) are in accordance with results of present study.

The viscosity values of unclarified acidic whey (shrikhand whey) beverage are in close agreement with the results of Sikder *et al.* (2001).

4.5.2 Proximate composition of acidic whey (shrikhand whey) beverage

The proximate composition of unclarified mango flavoured, prefiltered orange flavoured and ultrafiltered pineapple flavoured acidic whey (shrikhand whey) beverage is depicted in Table 10.

Table 10. Proximate composition of selected acidic whey (shrikhand whey) beverage

Chemical constituents (%)	Unclarified	Clarified	
	A	B	C
Moisture	82.50	83.60	84.10
Total solids	17.50	16.40	15.90
Fat	0.24	---	---
Protein	2.80	1.25	0.25
Lactose	3.60	3.38	3.20
Ash	0.60	0.55	0.42
Sugar	12.00	12.00	12.00
Acidity	0.40	0.33	0.31
pH	4.70	4.48	4.46

(Figure in parenthesis indicate mean value)

It can be noted from Table 10 that the unclarified mango flavoured acidic whey (shrikhand whey) beverage contained 17.50 per cent total solids i.e. highest compared to that of prefiltered orange flavoured and ultrafiltered pineapple flavoured acidic whey (shrikhand whey) beverage. It is due to higher protein and fat content in unclarified mango beverage.

Table 10 further indicated that the protein content of the unclarified mango flavoured acidic whey (shrikhand whey) beverage was higher i.e. 2.80 per cent than that of prefiltered orange flavoured and ultrafiltered pineapple flavoured acidic whey (shrikhand whey) beverage. It is because of the deproteinization operation carried out during prefiltration of ultrafiltration process.

It can be further observed from Table 10 that fat content of unclarified mango flavoured acidic whey (shrikhand whey) beverage was only 0.24 per cent but there was on fat in prefiltered and ultrafiltered beverage. It is because of defatting of the acidic whey (shrikhand whey) and process of filteration which removed the fat to a greater extent.

It can be observed from the Table 10 that the acidity of unclarified and clarified acidic whey (shrikhand whey) beverage was in the range of

0.40 to 0.31 per cent with the corresponding pH in the range of 4.70 to 4.46.

Table 10 indicated that there is little variation in lactose content of unclarified mango flavoured, prefiltered orange flavoured and ultrafiltered pineapple flavoured acidic whey (shrikhand whey) beverage. The ash content of the beverage also varies accordingly.

The results of the ultrafiltered, pineapple flavoured acidic whey (shrikhand whey) beverage in present study are well comparable with those reported by Suresha and Jayaprakash (2003). They reported that total solids, total protein, lactose and ash content of pineapple flavoured ultrafiltered whey permeate beverage were 15.97, 0.28, 5.01 and 0.56 per cent respectively.

The results of unclarified mango flavoured acidic whey (shrikhand whey) beverage in present study are in accordance with Sikder et al. (2001).

4.5.3 Nutritional quality of acidic whey (shrikhand whey) beverage

The nutritional quality of unclarified mango flavoured, prefiltered orange flavoured and ultrafiltered pineapple flavoured acidic whey (shrikhand whey) beverage is depicted in Table 11.

Table 11. Nutritional quality of selected acidic whey (shrikhand whey) beverage

Chemical constituents (%)	Unclarified	Clarified	
	A	B	C
Calcium (ppm)	359.2	360.05	401.80
Phosphorus (mg/100g)	250.0	250.00	255.00
Vit. C (mg/100g)	9.00	17.83	10.85
Vit. A (IU/100g)	727.15	393.164	686.785

(Figure in parenthesis indicates mean value)

The results in Table 11 shows that the calcium content of ultrafiltered pineapple flavoured acidic whey (shrikhand whey) beverage was highest i.e. 401.8 ppm followed by prefiltered orange flavoured acidic whey (shrikhand whey) beverage and unclarified mango flavoured acidic whey (shrikhand whey) beverage i.e. 360.5 mm and 359.2 ppm respectively.

It is revealed from Table 11 that the phosphorus content of ultrafiltered pineapple flavoured acidic whey (shrikhand whey) beverage was highest i.e. 255 mg/100g followed by prefiltered orange flavoured acidic whey (shrikhand whey) beverage and unclarified mango flavoured acidic whey (shrikhand whey) beverage i.e. 250 mg/100g and 250 mg/100, respectively.

The ascorbic acid content of prefiltered orange flavoured acidic whey (shrikhand whey) beverage was higher than that of other beverages i.e. 17.83 mg/100 g. The high value of ascorbic acid is mainly due to the higher content of ascorbic acid in orange juice as compared to that of mango and pineapple juice.

From Table 11, it is revealed that the vitamin A content of unclarified mango flavoured acidic whey (shrikhand whey) beverage was higher i.e. 727.15 IU/100 g followed by ultrafiltered pineapple flavoured acidic whey (shrikhand whey) and prefiltered orange flavoured beverage i.e. 686.785 IU/100 g and 393.164 IU/100 g respectively.

4.6 Carbonation of acidic whey (shrikhand whey) beverage

The selected unclarified mango flavoured (A), prefiltered orange flavoured (B) and ultrafiltered pineapple flavoured (C) acidic whey (shrikhand whey) beverages were carbonated at a pressure of 25, 30, 35 psi at 4 ± 1°C temperature. The effect of levels of carbonation on sensory quality of these beverages is depicted in Table 12.

The data in Table 12 indicates that effect of carbonation on all sensory parameters of beverage except colour was significant. It can also be seen from Table 12 that score for appearance parameter of sensory quality is higher for ultrafiltered pineapple flavoured acidic whey (shrikhand whey) beverage at 30 psi than unclarified mango flavoured and prefiltered orange flavoured acidic whey (shrikhand whey) beverage at 25 psi and 35 psi respectively. It is due to the tingling air bubbles in ultrafiltered pineapple flavoured acidic whey (shrikhand whey) beverage which add to the sparkling appearance.

From Table 12 it could be observed that the taste of carbonated unclarified mango flavoured acidic whey (shrikhand whey) beverage at 25 psi pressure level was higher whereas it goes on decreasing as the pressure level was increased. It further indicates that prefiltered orange flavoured and ultrafiltered pineapple flavoured acidic whey (shrikhand whey) beverage obtained higher score for taste parameter at 35 psi and 30 psi respectively. It is due to the tingling taste at these pressure levels which was 'liked extremely' by the judges.

Table 12. Effect of level of carbonation on sensory quality of selected acidic whey (shrikhand whey) beverage

Acidic whey (shrikhand whey) beverage	Carbonation level (psi)	Appearance (%)	Colour	Taste	Aroma	Consistency	Mouthfeel	Overall acceptability
A	25	8.06	8.20	8.16	8.10	8.20	8.36	8.33
	30	8.06	8.20	7.06	6.93	6.96	7.10	7.06
	35	7.86	8.20	6.83	6.96	6.86	6.83	6.73
B	25	8.00	8.03	7.13	7.03	6.56	6.93	6.86
	30	8.00	8.03	7.50	7.53	7.73	7.76	7.56
	35	8.06	8.03	8.13	8.20	8.10	8.00	8.10
C	25	7.90	7.80	7.66	7.00	7.53	7.86	7.83
	30	8.33	8.30	8.26	8.16	8.33	8.53	8.53
	35	7.90	7.86	7.70	7.53	7.53	7.93	7.90
SE	--	0.08	--	0.12	0.09	0.11	0.05	0.05
CD @ 5%	--	0.25	--	0.35	0.27	0.34	0.15	0.17

(Each value is the mean score of six judges in three trials)

59

It can be further stated from Table 12 that amongst all the various acidic whey (shrikhand whey) beverages prepared, the highly acceptable acidic whey (shrikhand whey) beverage was ultrafiltered pineapple flavoured acidic whey (shrikhand whey) beverage with a highest score of 8.53 at a pressure of 30 psi.

The results of the present study are in close agreement with those of the results reported by Suresha and Jayaprakasha (2003).

SUMMARY AND CONCLUSION

The present investigation was undertaken to utilize the acidic whey (shrikhand whey), a by-product of the dairy industry for the preparation of beverages considering the availability in large amount and nutritional quality of it. Attempts have been made to develop a process for preparation of carbonated acidic whey (shrikhand whey) beverage by blending with different types of fruit juices at various levels of concentration, which will improve the palatability and quality of acidic whey (shrikhand whey). The product prepared was organoleptically evaluated and its Physico-chemical and nutritional qualities were also studied.

The whey used in the preparation of beverage was obtained during the manufacturing of shrikhand by using 6 per cent standardized fat buffalo milk. The acidic whey (shrikhand whey) was passed through cream separator for the removal of residual fat. Then it was kept undisturbed till the acidity of 0.8 per cent was obtained. Later on, it was clarified by prefiltration and ultra filtration process. Prefiltration was carried out by passing through 'Microfiber Glassfilter and Mixed Esters of Cellulose' by applying vaccum. Then, it was ultrafiltered in 'Millipore Stirred Cell Ultrafiltration Unit' by using 300 KD 'Polyether Sulphone Biomax Ultrafiltration Disc'. This resulted into prefiltered and ultrafiltered acidic whey (shrikhand whey).

The Physico-chemical qualities of unclarified, prefiltered and ultrafiltered acidic whey (shrikhand whey) were studied and used for preparation of beverage base. The pasteurized beverage base was standardized by using sugar levels of 8.0, 10.0, 12.0, 14.0 per cent and 0.8 per cent acidity. Then, subjected to sensory evaluation to know the most acceptable level of sugar and it was found that 12.0 % sugar level beverage base was highly acceptable in all types of whey beverages.

The beverages adjudged as a best were subjected to Physico-chemical and nutritional quality assessment. The physical quality was assessed with reference to colour, specific gravity, viscosity, sedimentation and turbidity. The proximate compositions of whey beverages were determined with respect to moisture, total solids, fat, protein, lactose, titratable acidity, pH and ash.

The acidic whey (shrikhand whey) beverage base was blended with three different types of fruit juices i.e. mango, orange and pineapple at different levels of concentrations i.e. 18.0, 20.0, 22.0 and 24.0 per cent. Then, it was heated and homogenized. The product adjudged as best from each fruit juice were utilized for further study. The unclarified acidic whey (shrikhand whey) beverage base with 22 per cent unclarified mango juice concentration scored higher i.e. 8.23 for overall acceptability over other treatment. Whereas, prefiltered acidic whey (shrikhand whey) beverage base with 22 per cent orange juice concentration scored higher i.e. 7.96 for overall acceptability. The ultrafiltered acidic whey (shrikhand whey) beverage base with 22 per cent pineapple juice concentration scored 8.26 i.e. highest score and were found to be superior amongst all.

Nutritional quality of beverage was assessed with respect to calcium, phosphorus, Vitamin A and Vitamin C content.

The above selected beverages were carbonated at three different levels of carbonation i.e. 25, 30, 35 psi at $4 \pm 1°C$ temperature and were subjected to organoleptic evaluation. The unclarified mango beverage scored higher i.e. 8.33 at 25 psi. The prefiltered orange flavoured beverage scored higher i.e. 8.10 at 35 psi. The ultrafiltered pineapple flavoured beverage scored highest i.e. 8.53 at 30 psi, for overall acceptability than that of unclarified mango flavoured and prefiltered orange flavoured acidic whey (shrikhand whey) beverages.

CONCLUSION

It may be concluded that the process developed for preparation of carbonated fruit flavoured acidic whey (shrikhand whey) beverage was found to be highly acceptable considering the organoleptic, Physico-chemical and nutritional qualities. The pretreatments given for clarification improved physical qualities. The fruit juices incorporated helped in improving the organoleptic quality with respect to colour, flavour, taste, aroma and mouth feel. In addition to that, the Physico-chemical and nutritional qualities of the final product were also improved. Thus, acceptable and good quality acidic whey (shrikhand whey) beverage can be prepared with addition of 12.0 per cent sugar and 22.0 per cent pineapple juice carbonating at the pressure of 30 psi of carbon dioxide gas.

REFERENCES

1. Albrecht, P. (1986). Possibilities of effective utilization of whey. Prumysl Potravin, **37** (12):641-642.

2. Aneja, R.P. (1997). Traditional dairy delicacies. A compendium in Dairy India. 5th Edn. New Delhi, 371-386.

3. Anonymous (1982). Fruit whey beverage in small packs. Dairy Sci. Abstracts, 44.

4. A.O.A.C. (1990). Official methods of analysis. Association of Official Analytical Chemists, 14th Edn., Washington DC.

5. Balsubramanyam, B.V., Sudhir Singh and Bhanamurti, J.L. (1989). Precipitation of solids in whey from different sources. Indian J. Dairy Sci., **42** (2):301-304.

6. Bambha, P.P., Setty, A.A.S. and Namburipad, V.K. (1972). 'Whevit', a nourishing soft drink. Indian Dairyman, **24** (7):153-157.

7. Barbier, V.K. and Rialand, M.R. (1987). Preparation of fruit flavoured beverage using acid whey. Dairy Sci. Abstracts, **47**:129.

8. Belhe, N.D., Thorat, A.K., Kulkarni, M.B. and Salunkhe, D.K. (1982). Utilization of whey in bakery products. Dairy Guide, **4** (10):49-52.

9. Choi, H.S. and Kosikowski, F.V. (1985). Sweetened plain and flavoured carbonated yoghurt beverages. J. Diary Sci., **68**(3):613-619.

10. De Sukumar (1974). Outlines of dairy technology, Nineteenth Edn., Oxford University Press. 468-469.

11. Dhaka, V., Dabur, R.S. and Jyotika (2002). Utilization of whey for preparation of coffee drink. Indian J. Dairy Sci., **55** (2):86-88.

12. Durham, R.J., Howrigan, J.A., Sleigh, R.W. and Johnson, R.L. (1997). Whey fractionation. Food Australia, **49**(10):460-465.

13. F.A.O. (1995). Production year book. Food and Agricultural Organization of the United Nations, Rome, Italy.

14. Fauquant, J., Pierre, A. and Brule, G. (1985). Clarification of acid casein whey. Technique and marketing No. 1003, **41**:37-39.

15. Fresnel, J.M. and Moore, K.K. (1978). Swiss scientists develop soft drink form whey. Food Product Development, **12**(1):45.

64

16. Gagrani, R.L., Rathi, S.D. and Ingle, U.M. (1987). Preparation of fruit flavoured beverage from whey. J. of Food Sci. and Tech., **24** (2):93-94.

17. Ganasekar, R. and Balaraman, N. (2001). Utilization of whey in dairy rations. Indian J. Dairy Sci., **54**(3):118-128.

18. Gandhi, D.N. (1989). Production of some useful products of industrial importance through microbial fermentation of whey. Indian Dairyman, **41**(4):182-184.

19. Ghosh, S., Kanawjia, S.K. and Singh, S. (1995). Recent advances in whey permeate based sports drink. Beverage and Food World, **9**:24-26.

20. G.O.I. (2002). Economic survey 2001-2002. Economic Division, Ministry of Finance, Government of India, New Delhi.

21. Gupta, S.K. (1976). Sensory evaluation in food industry. Indian Dairyman, **28**(7):293-295.

22. Gupta, V.K. and Mathur, B.N. (1989). Current trends in whey utilization. Indian Dairyman, **41**(3):165-167.

23. Gupta, V.K. (2000). Overview of processing and utilization of dairy by products. Indian Dairyman, **52**(5):55-60.

24. Hofer, A. (1995). Whey a by-product or a source of valuable milk constituents. DMZ, Milchewirt Schaft, **116**:124-129.

25. Holsinger, V.H., Posati, L.P. and De Vilbiss, E.D. (1974). Whey beverages - A review. J. Dairy Sci., **57**(7):849.

26. I.S.I. (1981). IS : SP : 18 (Part XI). Handbook of food analysis. Part XI. Dairy Products. Indian Standards Institutions, Manak Bhavan, New Delhi.

27. Jadhav, P.E., Kulkarni, M.B. and Narwade, V.S. (1991). Factors in fluencing mineral content of chakka whey. Indian J. Dairy Sci., **44**(8): 510-513.

28. Jandal, J.M. (1996). Dairy beverages. Beverages and Food World, **11**:30-32.

29. Jayaprakasha, H.M., Anantakrishna, C.P., Atmaram, K. and Natrajan, A.M. (1985). Variation in yield, pH, and composition of whey used in soft drink. Cheiron, **14** (5):282-284.

30. Jayaprakasha, H.M., Anantakrishna, C.P., Atmaram, K. and Natrajan, A.M. (1986). Preparation of soft drink from clarified and deproteinized whey. Cheiron, 15(1):16-19.

31. Jayaprakasha, H.M., Anantakrishna, C.P.; Atmaram, K. and Natrajan, A.M. (1992). Preparation of soft drink from clarified and deproteinized whey. Indian J. Dairy Sci., 21(2):17-20.

32. Jelen, P. (1992). Whey cheeses and beverages. In whey and lactose processing. J. Zadow (ed.), Elserier Applied Science, London and New York.

33. Kanwar, S. and Chopra, D.K. (1976). Practical Agricultural Chemistry. Second Edition, ICAR, New Delhi.

34. Karnakar, S., Rao, B.V.R. and Rao, J.J. (1984). The preservative effect of solid carbon di oxide and gas on quality of salted and sweetened butter milk. J. Food Sci. Tech., 21:46-50.

35. Kar, T. and Misra, A.K. (1998). Utilization of chhana whey for alcohol production. Indian J. Dairy Sci., 51 (3): 204-206.

36. Kaur, P., Grewal, K.S. and Bakshi, A.K. (2000). Technology of whey based carrot juice beverage. Beverages and Food World, 3:19-20.

37. Khalikar, S.M. (1990). Studies on utilization of sour whey for preparation of soft beverage. M.Sc. Thesis, M.A.U., Parbhani (India.).

38. Khamrui, K. and Rajorhia, G.S. (1998). Formulation of RTS – whey based kinnow juice beverage. Indian J. Dairy Sci., 51(6):413-419.

39. Khurdia, D.S. (1989). Carbonation in fruit beverages. Beverage and Food World, April /June; 9-15.

40. Khurdia, D.S. (1990). A study on fruit juice based carbonated drink. Indian Food Packer. Nov/Dec., 45-50.

41. Kosikowski, F.V. (1968). Nutritional beverages from acid whey powder. J. Dairy Sci., 51 (8):1299-1301.

42. Kosikowski, F.V. (1979). Whey utilization and whey products. J. Dairy Sci., 62(7):1149-1151.

43. Krishnaiah, N., Reddy, C.R. and Ramarao, M. (1989). Studies on keeping quality of whey beverage. Asian J. Dairy Res., 8(1):8-14.

44. Krishnaiah, N., Reddy, C.R. and Rao, M.R. (1991). Development of beverage from acid whey. Indian J. Dairy Sci., 44:300-301.

45. Krishnaiah, N., Reddy, C.R. and Rao, M.R. (1998). Development of beverage from acid whey. Indian J. Dairy Sci., **46**:209-210.

46. Kulkarni, M.B., Chavan, I.G. and Belhe, N.D. (1987). Chemical composition of chakka whey. Indian J. Dairy Sci., **40** (1):65-68.

47. Kumar, A., Tiwarii, B.D. and Rai, V.K. (2003). Delicious soft drinks from whey. Beverage and Food World., **8**:39-43.

48. Mandal, P.L., Ghatak, P.K. and Bandopadhyay, A.K. (1997). Studies on shelf life of whey beverage. Indian J. Dairy Sci., **50**(3):193-198.

49. Marwaha, S.S. and Kenedy, J.F. (1988). Whey pollution problem and potential utilization. International J. Food Sci. and Tech., **23**:323-336.

50. Mathur, B.N., Abhay Kumar and Ladkani, B.G. (1986). Clarification of whey for the preparation of beverages. Indian J. Dairy Sci., **39**(3):340-342.

51. Mathur, B.N., Kumar, A. and Ladkani, B.G. (1988). UHT processed beverage paved way for economic utilization of whey. Indian Dairyman, **8**(10):533-535

52. Mohanty, A. K., Pattnaik, P., Mukhopadhyay, U. K, Grover, S. and Batish,.V.K. (1998). Value added products from whey. Beverage and Food World. **1**:19-21.

53. Niketic, G. and Marinkovik, S. (1984). Production of refreshing beverage from whey under aseptic conditions. Mijekarstvo, **34**(4):105-109.

54. Nelson, F.E., Brown, W.C. and Metawally, M.M. (1972). Whey as a component of fruit flavoured drinks. J. Dairy Sci., **54**(5):758.

55. Pagote, C.N. and Balachandran (1993). Directly acidified milk based soft drinks. Beverage and Food World. **11**:18-19.

56. Panse, V.G. and Sukhatme, P.V. (1985). Statistical methods for Agril. workers. ICAR Pub., 2nd Edn. New Delhi.

57. Patel, R.s., Jayaprakasha, H.M. and Singh, S. (1991). Recent advantages in concentration and drying of whey. Indian Dairyman : 417-421.

58. Paul, S.C. (1990). Nutritive beverages for product diversification in dairy industries. Indian J. Dairy Sci., **44** (8):282-287.

59. Paul, D., Mukhopadhyay, R., Chatarjee, B.P. and Guha, A.K. (2002). Lactose from whey evaluation of different isolation procedures on yield and quality. Indian J. Dairy Sci., **55**(2):65-67.

60. Prasad, K., Sharma, H.K., Mahajan, D. and Jaya (2001). Utilization of whey based mango beverage. Beverage and Food World, **11**:31-32.

61. Prendergast,K.(1985). Whey drinks- technology, processing and marketing.J.Society of Dairy Tech.,**38**(4):103-105

62. Rajeshkumar, Patil, G.R. and Rajor, R.B. (1987). Development of lassi type cultured beverage from cheese whey. Asian J. Dairy Res., **6**(3):121-124.

63. Raganna, S. (1986). Handbook of analysis and quality control for fruits and vegetable products, 2nd Edn. ICAR, New Delhi.

64. Reddy, G.J., Rao, B.V.R., Reddy, K.S.R. and Venkayya, D. (1987). Development of a whey beverage. Indian J. Dairy Sci., **40**(4):445-449.

65. Sangu, K.P.S. (2004). Development of RTS whey based hot soup. Indian J. Dairy Sci., **57**(2):94-95.

66. Sarvana, R. and Manimegalai, G. (2002a). Studies on whey based jack fruit RTS beverages. Beverage and Food World, **1**:57-58.

67. Sarvana, R. and Manimegalai, G. (2002b). A delicious soy, milk whey blended papaya RTS, **8**:42-43.

68. Sarvana, R. and Manimegalai, G. (2003).A study on storage behaviour of whey based pineapple juice RTS beverage. Indian Food Packer, **2**:51-55.

69. Schuster, J. (1977). Process for producing refreshing beverage free from alcohol CO_2 and preservative. Dairy Sci. Abstracts. 40.

70. Shaikh, S.Y., Rathi, S.D., Pawar, V.D. and Agarkar, B.S. (2001). Studies on development of a process for preparation of fermented carbonated whey beverage. J. Food Sci. and Tech., **38**(5):519-521.

71. Shilpa, Vij and Gandhi, D.N. (1993). Whey as alternative substrate for the production of baker's yeast. Indian Food Industry, **12**(5):41-43.

72. Shitova, L.A., Dechenka, V.V., Borovski, V.R., Yankovskaya, N.E., Shelimanov, V. A. and Borshch, G.G. (1991). Manufacture of dairy products. USSR patent SU 1634227.

73. Shukla, F.C., Sharma, A. and Baljit Singh (2004). Studies on the preparation of fruit beverages using whey and butter milk. J. Food Sci. Tech., **41**(1):102-105.

74. Sikder, B., Sarkar, K., Ray, P.R. and Ghatak, P.K. (2001). Studies on the shelf life of whey based mango beverage. Beverage and Food World, **10**:53-62.

75. Singh, G.P. and Mathur, M.L. (1973). Whey and its composition. Indian J. Dairy Sci., **24**(3):201-202.

76. Singh, S., Ladkani, B.G., Kumar, A. and Mathur, B.N. (1994). Development of whey based beverages. Indian J. Dairy Sci., **47**(7):585-590.

77. Singh, S., Singh, A.K. and Patil, G.R. (2002). Whey utilization for health beverages. Indian Food Industry, **21**(4):38-41.

78. Skudra, L. and Reinikora, E. (1976). Effect of carbondioxide concentration on the micro flora in butter milk cocktails, Lativijas. Lauksaimniecibas Akademijas No. **88**, pp:3-6.

79. Suresha, K.B. and Jayaprakash, H.M. (2003). Utilization of ultrafiltration whey permeate for preparation of beverage. Indian J. Dairy Sci., **56**(5): 278-284.

80. Suresha, K.B. and Jayaprakash, H.M. (2004). Process optimization for preparation of beverage from lactose hydrolyzed whey permeate. J. Food Sci. and Tech., **41**(1) :27-32.

81. Surzanne Nielsen, S. (1992). Introduction to the chemical analysis of foods. Jones and Bartlett Publishers, Boston, U.K. : 253-254.

82. Tondon, H.L.S. (1993). Methods of analysis of solis, plants, waters, fertilizers. Fertilizers development and consultation organization (FDCO), New Delhi.

83. Wazirsingh, Kapoor, C.M. and Shrivastava, D.N. (1999). Standardization of technology for the manufacture of guava whey beverage. Indian J. Dairy Sci., **52**(5):268-271.

CARBONATED WHEY BEVERAGES

By
Katke S.D. and Patil P.S.

Rs: 700 /-

CARBONATED WHEY BEVERAGES

Katke S. D. and Patil P. S.

© 2017 by Research maGma Book Publication
All rights reserved. No part of this publication may be reproduced or transmitted, in any form or by any means, without prior permission of the author. Any person who does any unauthorized act in relation to this publication may be liable to criminal prosecution and civil claims for damages.
[The responsibility for the facts stated, conclusions reached, etc., is entirely that of the author. The publisher is not responsible for them, whatsoever.]

ISBN - 978-1-387-00252-8

Published & Printed By,
Research maGma Book Publication
HON- 14/87, Akkalkot Road, Gandhi Nagar,
Solapur, Maharashtra, India.
Contact No. : +91 7385878362
Website : http://researchmagma.com
Email ID: info@researchmagma.com

DEDICATION

I dedicate this book to my beloved parents (Mr. Dilipkumar Katke & Mrs. Rekha Katke) & my wife (Mrs. Puja Katke) as they are the driving forces throughout my life & career.

(Katke S.D.)

I dedicate this book to my beloved parents (Mr. Shashikant Patil & Mrs. Vidya Patil) & my husband (Mr. Sandip Patil) as they are the driving forces throughout my life & career.

(Patil P.S.)

PREFACE TO THE FIRST EDITION

For many years, an acute need has existed for carbonated whey beverage textbook that is suitable for Food Technology / Dairy Technology students with basic concepts. This book is designed primarily to fill the aforementioned need, and secondarily, to serve as a reference source for persons involved in food research, food product development, quality assurance, food processing, and in other activities related to the food & dairy industry. The original idea in the preparation of this book was to present basic information on the preparation of carbonated whey beverages. The basic principles remain the same, but much additional research carried out in recent years has extended and deepened our knowledge.

I believe the end product, considering it is a first edition, is really quite satisfying except perhaps for the somewhat generous length. If the readers concur with my judgment, I will be pleased. Organization of the book is quite simple and I hope appropriate.

Complete coverage of all aspects of carbonated beverages, of course, has not been attempted. It is hoped, however, that the topics of greatest importance have been treated adequately. In order to help & achieve this objective, emphasis has been given to broadly based principles.

Figures and tables have been used liberally in the belief that this approach facilitates understanding of the subject matter presented.

The number of references cited should be adequate to permit easy access to additional information.

To all readers, I extend an invitation to report errors that no doubt have escaped my attention, and to offer suggestions for improvements that can be incorporated in future (hopefully) editions.

Since enjoyment is an unlikely reader response to this book, the best I can hope for is that readers will find it enlightening and well suited for its intended purpose.

Katke S.D.
Patil P.S.

FOREWORD

India is an agricultural nation. It has nearly a tenth of world's arable land & a fifth of irrigated land. It has largest cattle population, second largest goat & sheep population & a vast coastline. India has potential for food processing due to strong demand of processed food products. India ranks No.01 in the world in the production of milk. Milk has been rated as complete food for infants and a rich food supplement for the adults. It has been also recognized as the excellent food for the maintenance of the health and promotion of the growth of the human beings. In India, the milk is consumed in the form of the fluid milk and various milk products i.e. butter, ghee, khoa, cheese, paneer, shrikhand, ice-cream etc. During manufacture of certain milk products, the by-products gets produced that too in very large amount than the products manufactured. The major by-products of the dairy industry are skim milk, butter milk and whey. Now a day's more emphasis is given for economic utilization of these by-products as they contain valuable milk solids.

The utilization of these by-products has not only increased the availability of nutritional foods but has also indicated the profitable method of their utilization with minimizing the problem of pollution. In India, whey is obtained as the by-product in the preparation of chhana, paneer, cheese, casein and shrikhand. The whey which is used as the waste effluent could be used in the formulation of nutritious, palatable and therapeutic beverages.

I believe that this book will serve as a boon for the beginners to enhance their knowledge about whey beverage processing. Authors took every possible effort for a systematic presentation of content. My best wishes to all for their academic endeavor.

Dr. Rodge A.B.
Principal, MGM College of Food Technology,
Gandheli, Aurangabad, Maharashtra.

www.ingramcontent.com/pod-product-compliance
Lightning Source LLC
Chambersburg PA
CBHW021901170526
45157CB00005B/1917